THE AMERICAN REVOLUTION

BY

JOHN FISKE

IN TWO VOLUMES

VOL. I.

BOSTON AND NEW YORK
HOUGHTON, MIFFLIN AND COMPANY
The Riverside Press, Cambridge
1896

ELEVENTH THOUSAND

The Riverside Press, Cambridge, Mass., U. S. A.
Electrotyped and Printed by H. O. Houghton & Co.

To

MRS. MARY HEMENWAY

IN RECOGNITION OF THE RARE FORESIGHT AND PUBLIC SPIRIT WHICH SAVED FROM DESTRUCTION ONE OF THE NOBLEST HISTORIC BUILDINGS IN AMERICA, AND MADE IT A CENTRE FOR THE TEACHING OF AMERICAN HISTORY AND THE PRINCIPLES OF GOOD CITIZENSHIP

I DEDICATE THIS BOOK

PREFACE.

In the course of my work as assistant librarian of Harvard University, about seventeen years ago, I had occasion to overhaul what we used to call the "American Room," and to superintend or revise the cataloguing of some thousands of titles of volumes and pamphlets relating to America. In the course of this work my attention was called more and more to sundry problems and speculations connected with the transplantation of European communities to American soil, their development under the new conditions, and the effect of all this upon the general progress of civilization. The study of aboriginal America itself presented many other interesting problems. In 1879, 1881, and 1882, I gave courses of lectures at the Old South Meeting-House in Boston, in aid of the fund for the preservation of that venerable building, and in pursuance of Mrs. Hemenway's scheme for making it a place for the teaching of American history. As to the success of that scheme we may now speak with some satisfaction. The preservation of the noble old church may be regarded as assured; the courses of instruction there given in American history and cognate subjects are attended by thousands, old and young, especially by school-teachers and their pupils; and similar courses of study have already been inaugurated in several other cities and towns. It is believed that the good results of this work will be manifold.

As regards my lectures, just mentioned, they dealt chiefly with the discovery and colonization of America, and contained sundry generalizations since embodied in "American Political Ideas" and in the first chapter of "The Beginnings of New England." Some further generalizations of a similar sort will be worked out in my forthcoming book — now in press — "The Discovery of America."

While busy in this work, the plan occurred to me in 1881 of writing a narrative history of the United States, neither too long to be manageable nor too brief to be interesting, something that might comprise the whole story from 1492 to (say) 1865 within four octavos, like the book of my lamented friend, the late John Richard Green. Plans of this sort, to be properly carried out, require much time, and a concurrence of favourable circumstances, as Mr. Cotter Morison has pointed out in his sketch of Gibbon. If my plan is ever fully realized, it can only be after many years.

Meanwhile it has seemed to me that fragments of the work might as well be published from time to time as to be lying idle in manuscript in a cupboard. It was with this feeling that "The Critical Period of American History" and "The Beginnings of New England" were brought out, and it is with the same feeling that these volumes on "The American Revolution" are now offered to the public.

In writing the story of this period my design was not so much to contribute new facts as to shape the narrative in such a way as to emphasize relations of cause and effect that are often buried in the mass of details. One is constantly tempted, in such a narrative, to pause for discussion, and to add item upon item of circumstantial description because it is interesting in itself; but in conformity with the plan of the book of which this was to have been a part, it was necessary to withstand such temptations. I have not even undertaken to mention all the events of the Revolutionary War. For example, nothing is said about the Penobscot expedition, which was a matter of interest to the people of Massachusetts, but of no significance in relation to the general history of the war.

The present work is in no sense "based upon" lectures, but it has been used as a basis for lectures. When I had nearly finished writing it, in 1883, I happened to read a few passages to some friends, and was thereupon urged to read the whole work, or the greater part of it, as lectures.

This was done in the Old South Meeting-House early in 1884. The lectures were afterward given in many towns and cities, from Maine to Oregon, usually to very large audiences. In Boston, New York, and St. Louis the whole course was given from two to five times; and single lectures were repeated in many places. I was greatly surprised at the interest thus shown in a plain narrative of events already well known, and have never to this day understood the secret of it.

On some accounts I should have been glad to withhold this book some years longer, in the hope of changing its plan somewhat and giving the subject a fuller treatment, now that it is not to appear as part of a larger work. But so many requests have been made for the story in book form that it has seemed best to yield to them. In relation to these two volumes, "The Critical Period of American History" now stands as a third volume. The narrative is continuous from the one to the other.

I have not thought it worth while to add to the present work a bibliographical note, because, in view of the existence of Mr. Justin Winsor's "Reader's Handbook of the American Revolution," such a note would be quite superfluous. Mr. Winsor's book contains a vast amount of bibliographical information, most lucidly arranged, within a very small compass, and costs but a trifle. From it the general reader can find out "where to go" for further information concerning any and all points that may come up in these volumes; and if then he still wants more, he may consult the sixth and seventh volumes of Winsor's "Narrative and Critical History of America."

The portrait of Washington prefixed to this volume is now engraved, I believe, for the first time. It is from a miniature enamelled on copper by Henry Bone, R. A., an artist preëminent for skill in such work. Bone appears to have followed an original crayon sketch of Washington made in 1796 by William Birch, to whom he has given credit by putting his initials, W. B., on the miniature. This has given rise to an error; the initials have been mistaken for those of Bone himself, who has thus been wrongly spoken of as "William" Bone. The miniature was made for a family in England. After some years this family became straitened in circumstances, and the miniature was bought by George Peabody, who gave it to a lady in London interested in such things. By 1870 this lady in turn became desirous of obtaining its value, and a communication upon the subject was published in the Proceedings of the Massachusetts Historical Society for April of that year. It has since then come into the hands of Mrs. Hemenway of Boston, its present owner.

My thanks are due to Messrs. G. P. Putnam's Sons for permission to use the map of the battle of Monmouth.

St. Louis, *April* 14, 1891.

CONTENTS.

CHAPTER I.

THE BEGINNINGS.

CHAPTER II.

THE CRISIS.

CHAPTER III.

THE CONTINENTAL CONGRESS.

CHAPTER IV.

INDEPENDENCE.

CHAPTER V.

FIRST BLOW AT THE CENTRE.

CHAPTER VI.

SECOND BLOW AT THE CENTRE.

CHAPTER VII.

SARATOGA.

PORTRAIT AND MAPS.

THE AMERICAN REVOLUTION.

CHAPTER I.

THE BEGINNINGS.

DURING the seventy years which elapsed between the overthrow of the Stuart dynasty and the victory of Wolfe on the Heights of Abraham, the relations between the American colonies and the British government were, on the whole, peaceful; and the history of the colonies, except for the great and romantic struggle with New France, would have been almost destitute of striking incidents. In view of the perpetual menace from France, it was manifestly unwise for the British government to irritate the colonies, or do anything to weaken their loyalty; and they were accordingly left very much to themselves. Still, they were not likely to be treated with any great liberality, — for such was not then, as it is hardly even yet, the way of governments, — and if their attachment to England still continued strong, it was in spite of the general demeanour of the mother-country. Since 1675 the general supervision of the colonies had been in the hands of a standing committee of the Privy Council, styled the "Lords of the Committee of Trade and Plantations," and familiarly known as the "Lords of Trade." To this board the governors sent frequent and full reports of the proceedings in the colonial legislatures, of the state of agriculture and trade, of the revenues of the colonies, and of the way in which the public money was spent. In private letters, too, the governors poured forth their complaints into the ears of the Lords of Trade, and these complaints were many and loud. Except in Pennsylvania and Maryland, which were like hereditary monarchies, and in Connecticut and Rhode Island, where the governors were elected by the people, the colonial governors were now invariably appointed by the Crown. In most cases they were inclined to take high views regarding the royal prerogative, and in nearly all cases they were unable to understand the political attitude of the colonists, who on the one hand gloried in their connection with England, and on the other hand, precisely because they were Englishmen, were unwilling to yield on any occasion whatsoever one jot or tittle of their ancient liberties. Moreover, through the ubiquity of the popular assemblies and the directness of their control over the administration of public affairs, the political life of America was both really and ostensibly freer than that of England was at that time; and the ancient liberties of Englishmen, if not better preserved, were at least more conspicuously asserted. As a natural consequence, the royal governors were continually trying to do things which the people would not let them do, they were in a chronic state of angry warfare with their assemblies, and they were incessant in their com-

The Lords of Trade.

plaints to the Lords of Trade. They represented the Americans as factious and turbulent people, with their heads turned by queer political crotchets, unwilling to obey the laws, and eager to break off their connection with the British Empire. In this way they did much to arouse an unfriendly feeling toward the colonies, although eminent Englishmen were not wanting who understood American affairs too well to let their opinions be thus lightly influenced. Upon the Lords of Trade these misrepresentations wrought with so much effect that now and then they would send out instructions to suspend the writ of *habeas corpus*, or to abridge the the freedom of the press. Sometimes their acts were absurdly arbitrary. In New Hampshire, the people maintained that as free-born Englishmen they had the right to choose their representatives; but the governor held, on the contrary, that this was no right, but only a privilege, which the Crown might withhold, or grant, or revoke, all at its own good pleasure. To uphold the royal prerogative, the governor was instructed to issue writs for elections to some of the towns, while withholding them from others; but the resistance of the people to this piece of tyranny was so determined that the Lords of Trade thought it best to yield. In Massachusetts, for more than thirty years, there went on an unceasing controversy between the General Court and the successive royal governors, Shute, Burnet, and Belcher, with reference to the governor's salary. The Lords of Trade insisted that the governor should be paid a fixed salary; but lest this should make the governor too

The governor's salary.

independent, the General Court obstinately refused to establish a salary, but made grants to the governor from year to year, in imitation of the time-honoured usage of Parliament. This method was, no doubt, inconvenient for the governors; but the colonists rightly valued it as one of the safeguards of popular liberty, and to their persistent refusal the Crown was obliged to give way. Similar controversies, in New York and South Carolina, were attended with similar results; while in Virginia the assembly more than once refused to vote supplies, on the ground that the liberties of the colony were in danger.

Such grievances as these, reported year by year to the Lords of Trade, and losing nothing in the manner in which they were told, went far to create in England an opinion that America was a lawless country, and sorely in need of a strong government. From time to time various schemes were proposed for limiting the powers of the colonial assemblies, for increasing the power of the governors, for introducing a titled nobility, for taxing the colonists by act of Parliament, or for weakening the feeling of local independence by uniting several colonies into one. Until after the French troubles had been disposed of, little came of any of these schemes. A plan for taxing the colonies was once proposed to Sir Robert Walpole, but the sagacious old statesman dismissed it with a laugh. "What!" said he. "I have half of Old England set against me already, and do you think I will have all New England likewise?" From time to time the liberal charters of Rhode

Sir Robert Walpole.

Island and Connecticut were threatened, but nothing came of this. But in one direction the Lords of Trade were more active. One of their most cherished plans was to bring about a union of all the colonies under a single head; but this was not to be a union of the kind which the Americans, with consummate statesmanship, afterward wrought out for themselves. It was not to be a union based upon the idea of the sacredness of local self-government, but it was a union to be achieved, as far as possible, at the expense of local self-government. To bring all the colonies together under a single viceroy would, it was thought, diminish seriously the power of each local assembly, while at the same time such a union would no doubt make the military strength of the colonies much more available in case of war. In 1764, Francis Bernard, governor of Massachusetts, wrote that "to settle the American governments to the greatest possible advantage, it will be necessary to reduce the number of them; in some places to unite and consolidate; in others to separate and transfer; and in general to divide by natural boundaries instead of imaginary lines. If there should be but one form of government established for the North American provinces, it would greatly facilitate the reformation of them." As long ago as 1701, Robert Livingston of New York had made similar suggestions; and in 1752, Dinwiddie of Virginia recommended that the Northern and Southern colonies be united respectively into two great confederacies.

The desirableness of bringing about a union of the colonies was also recognized by all the most liberal-minded American statesmen, though from a very different point of view. They agreed with the royal governors and with the Lords of Trade as to the urgent need for concentrating the military strength of the colonies, and they thought that this end could best be subserved by some kind of federal union. But at the same time they held that the integrity of the local self-government of each colony was of the first importance, and that no system of federation would be practicable which should in any degree essentially impair that integrity. To bring about a federal union on such terms was no easy matter; it was a task fitted to tax the greatest of statesmen at any time. At that time it was undoubtedly a hopeless task. The need for union was not generally felt by the people. The sympathies between the different colonies were weak and liable to be overborne by prejudices arising from rivalry or from differences in social structure. To the merchant of Boston, the Virginian planter was still almost a foreigner, though both the one and the other were pure-blooded Englishmen. Commercial jealousies were very keen. Disputes about boundaries were not uncommon. In 1756, Georgia and South Carolina actually came to blows over the navigation of the Savannah river. Jeremiah Dummer, in his famous "Defence of the New Charters," said that it was impossible that the colonies should ever be brought to unite; and Burnaby thought that if the hand of Great Britain were once taken off, there would be chronic civil war all the way from Maine to Georgia.

Weakness of the sentiment of union.

In 1754, the prospect of immediate war with the French led several of the royal governors to call for a congress of all the colonies, to be held at Albany. The primary purpose of the meeting was to make sure of the friendship of the Six Nations, and to organize a general scheme of operations against the French. The secondary purpose was to prepare some plan of confederation which all the colonies might be persuaded to adopt. New Hampshire, Massachusetts, Rhode Island, Connecticut, New York, Pennsylvania, and Maryland — only seven colonies of the thirteen — sent commissioners to this congress. The people showed little interest in the movement. It does not appear that any public meetings were held in favour of it. Among the newspapers, the only one which warmly approved of it seems to have been the "Pennsylvania Gazette," edited by Benjamin Franklin, which appeared with a union device and the motto "Unite or Die!"

The Albany Congress.

The circumstances of Franklin's life, no less than the vast sweep of his intelligence, had fitted him for sounder and wider views of the political needs of the time than were taken by most of his contemporaries. In a certain sense he may be said to have belonged to two very different colonies; nor was he unfamiliar with ideas current in the mother-country. During the session of the Albany Congress, a first attempt was made to establish a permanent union of the thirteen colonies. It was to Franklin that the plan was chiefly due. The legislative assembly of each colony was to choose, once in three years, rep-

Franklin's plan of union, 1754.

resentatives to attend a federal Grand Council; which was to meet every year at Philadelphia, a town which could be reached by a twenty days' journey either from South Carolina or from New Hampshire. This Grand Council was to choose its own speaker, and could neither be dissolved nor prorogued, nor kept sitting longer than six weeks at any one time, except by its own consent or by especial order of the Crown. The Grand Council was to make treaties with the Indians and to regulate the Indian trade; and it was to have sole power of legislation on all matters concerning the colonies as a whole. To these ends, it could levy taxes, enlist soldiers, build forts, and nominate all civil officers. Its laws were to be submitted to the king for approval, and the royal veto, in order to be of effect, must be exercised within three years.

To this Grand Council each colony was to send a number of representatives, proportioned to its contributions to the continental military service; yet no colony was to send less than two or more than seven representatives. With the exception of such matters of general concern as were to be managed by the Grand Council, each colony was to retain its powers of legislation intact. On an emergency, any colony might singly defend itself against foreign attack, and the federal government was prohibited from impressing soldiers or seamen without the consent of the local legislature.

The supreme executive power was to be vested in a president or governor-general, appointed and paid by the Crown. He was to nominate all military officers, subject to the approval of the Grand

Council, and was to have a veto on all the acts of the Grand Council. No money could be issued save by joint order of the governor-general and the council.

This plan, said Franklin, "is not altogether to my mind, but it is as I could get it." It should be observed, to the credit of its great author, that this scheme, long afterward known as the "Albany Plan," contemplated the formation of a self-sustaining federal government, and not of a mere league. As Frothingham well says, "It designed to confer on the representatives of the people the power of making laws acting directly on individuals, and appointing officers to execute them, and yet not to interfere with the execution of the laws operating on the same individuals by the local officers." It would have erected "a public authority as obligatory in its sphere as the local governments were in their spheres." In this respect it was much more complete than the scheme of confederation agreed on in Congress in 1777, and it afforded a valuable precedent for the much more elaborate and perfect Federal Constitution of 1787. It was in its main features a noble scheme, and the great statesman who devised it was already looking forward to the immense growth of the American Union, though he had not yet foreseen the separation of the colonies from the mother-country. In less than a century, he said, the great country behind the Alleghanies must become "a populous and powerful dominion;" and he recommended that two new colonies should at once be founded in the West, — the one on Lake Erie, the other in the valley of the Ohio, — with free chartered governments like those of Rhode Island and Connecticut.

But public opinion was not yet ripe for the adoption of Franklin's bold and comprehensive ideas. Of the royal governors who were anxious to see the colonies united on any terms, none opposed the plan except Delancey of New York, who wished to reserve to the governors a veto upon all elections of representatives to the Grand Council. To this it was rightly objected that such a veto power would virtually destroy the freedom of elections, and make the Grand Council an assembly of creatures of the governors. On the popular side the objections were many. The New England delegates, on the whole, were the least disinclined to union; yet Connecticut urged that the veto power of the governor-general might prove ruinous to the whole scheme; that the concentration of all the military forces in his hands would be fraught with dangers to liberty; and that even the power of taxation, lodged in the hands of an assembly so remote from local interests, was hardly compatible with the preservation of the ancient rights of Englishmen. After long debate, the assembly at Albany decided to adopt Franklin's plan, and copies of it were sent to all the colonies for their consideration. But nowhere did it meet with approval. The mere fact that the royal governors were all in favour of it — though their advocacy was at present, no doubt, determined mainly by military reasons — was quite enough to create an insuper-

Rejection of the plan.

able prejudice against it on the part of the people. The Massachusetts legislature seems to have been the only one which gave it a respectful. consideration, albeit a large town meeting in Boston denounced it as subversive of liberty. Pennsylvania rejected it without a word of discussion. None of the assemblies favoured it. On the other hand, when sent over to England to be inspected by the Lords of Trade, it only irritated and disgusted them. As they truly said, it was a scheme of union "complete in itself;" and ever since the days of the New England confederacy the Crown had looked with extreme jealousy upon all attempts at concerted action among the colonies which did not originate with itself. Besides this, the Lords of Trade were now considering a plan of their own for remodelling the governments of the colonies, establishing a standing army, enforcing the navigation acts, and levying taxes by authority of Parliament. Accordingly little heed was paid to Franklin's ideas. Though the royal governors had approved the Albany plan, in default of any scheme of union more to their minds, they had no real sympathy with it. In 1756, Shirley wrote to the Lords of Trade, urging upon them the paramount necessity for a union of the American colonies, in order to withstand the French; while at the same time he disparaged Franklin's scheme, as containing principles of government unfit even for a single colony like Rhode Island, and much more unfit for a great American confederacy. The union, he urged, should be effected by act of Parliament, and

Shirley recommends a stamp act.

by the same authority a general fund should be raised to meet the expenses of the war, — an end which Shirley thought might be most speedily and quietly attained by means of a "stamp duty." As Shirley had been for fifteen years governor of Massachusetts, and was now commander-in-chief of all the troops in America, his opinion had great weight with the Lords of Trade; and the same views being reiterated by Dinwiddie of Virginia, Sharpe of Maryland, Hardy of New York, and other governors, the notion that Parliament must tax the Americans became deeply rooted in the British official mind.

Nothing was done, however, until the work of the French war had been accomplished. In 1761, it was decided to enforce the Navigation Act, and one of the revenue officers at Boston applied to the superior court for a "writ of assistance," or general search-warrant, to enable him to enter private houses and search for smuggled goods, but without specifying either houses or goods. Such general warrants had been allowed by a statute of the bad reign of Charles II., and a statute of William III., in general terms, had granted to revenue officers in America like powers to those they possessed in England. But James Otis showed that the issue of such writs was contrary to the whole spirit of the British constitution. To issue such universal warrants allowing the menials of the custom-house, on mere suspicion, and perhaps from motives of personal enmity, to invade the home of any citizen, without being held responsible for any rudeness they

Writs of assistance.

might commit there, — such, he said, was "a kind of power, the exercise of which cost one king of England his head and another his throne;" and he plainly declared that even an act of Parliament which should sanction so gross an infringement of the immemorial rights of Englishmen would be treated as null and void. Chief Justice Hutchinson granted the writs of assistance, and as an interpreter of the law he was doubtless right in so doing; but Otis's argument suggested the question whether Americans were bound to obey laws which they had no share in making, and his passionate eloquence made so great an impression upon the people that this scene in the court-room has been since remembered — and not unjustly — as the opening scene of the American Revolution.

In the same year the arbitrary temper of the government was exhibited in New York. Down to this time the chief justice of the colony had held office only during good behaviour, and had been liable to dismissal at the hands of the colonial assembly. The chief justice was now made removable only by the Crown, a measure which struck directly at the independent administration of justice in the colony. The assembly tried to protect itself by refusing to assign a fixed salary to the chief justice, whereupon the king ordered that the salary should be paid out of the quit-rents for the public lands. At the same time instructions were sent to all the royal governors to grant no judicial commissions for any other period than "during the king's pleasure;" and to show that this was meant in

The chief justice of New York.

earnest, the governor of New Jersey was next year peremptorily dismissed for commissioning a judge "during good behaviour."

In 1762, a question distinctly involving the right of the people to control the expenditure of their own money came up in Massachusetts. Governor Bernard, without authority from the assembly, had sent a couple of ships to the northward, to protect the fisheries against French privateers, and an expense of some £400 had been thus incurred. The assembly was now ordered to pay this sum, but it refused to do so. "It would be of little consequence to the people," said Otis, in the debate on the question, "whether they were subject to George or Louis, the king of Great Britain or the French king, if both were arbitrary, as both would be, if both could levy taxes without Parliament." A cry of "Treason!" from one of the less clear-headed members greeted this bold statement; and Otis, being afterward taken to task for his language, published a "Vindication," in which he maintained that the rights of a colonial assembly, as regarded the expenditure of public money, were as sacred as the rights of the House of Commons.

Otis's "Vindication."

In April, 1763, just three years after the accession of George III., George Grenville became Prime Minister of England, while at the same time Charles Townshend was First Lord of Trade. Townshend had paid considerable attention to American affairs, and was supposed to know more about them than any other man in England. But his studies had led him to the conclusion that the colonies ought to be deprived of their self-government, and that a standing army ought to be maintained in America by means of taxes arbitrarily assessed upon the people by Parliament. Grenville was far from approving of such extreme measures as these, but he thought that a tax ought to be imposed upon the colonies, in order to help defray the expenses of the French war. Yet in point of fact, as Franklin truly said, the colonies had "raised, paid, and clothed nearly twenty-five thousand men during the last war, — a number equal to those sent from Great Britain, and far beyond their proportion. They went deeply into debt in doing this; and all their estates and taxes are mortgaged for many years to come for discharging that debt." That the colonies had contributed more than an equitable share toward the expenses of the war, that their contributions had even been in excess of their ability, had been freely acknowledged by Parliament, which, on several occasions between 1756 and 1763, had voted large sums to be paid over to the colonies, in partial compensation for their excessive outlay. Parliament was therefore clearly estopped from making the defrayal of the war debt the occasion for imposing upon the colonies a tax of a new and strange character, and under circumstances which made the payment of such a tax seem equivalent to a surrender of their rights as free English communities. In March, 1764, Grenville introduced in the House of Commons a series of Declaratory Resolves, announcing the intention of the government to raise

Expenses of the French War.

Grenville's Resolves.

a revenue in America by requiring all legal documents to bear stamps, varying in price from threepence to ten pounds. A year was to elapse, however, before these resolutions should take effect in a formal enactment.

It marks the inferiority of the mother-country to the colonies in political development, at that time, that the only solicitude as yet entertained by the British official mind, with regard to this measure, seems to have been concerned with the question how far the Americans would be willing to part with their money. With the Americans it was as far as possible from being a question of pounds, shillings, and pence; but this was by no means correctly understood in England. The good Shirley, although he had lived so long in Massachusetts, had thought that a revenue might be most easily and quietly raised by means of a stamp duty. Of all kinds of direct tax, none, perhaps, is less annoying. But the position taken by the Americans had little to do with mere convenience; it rested from the outset upon the deepest foundations of political justice, and from this foothold neither threatening nor coaxing could stir it.

The first deliberate action with reference to the proposed Stamp Act was taken in the Boston town meeting in May, 1764. In this memorable town meeting Samuel Adams drew up a series of resolutions, which contained the first formal and public denial of the right of Parliament to tax the colonies without their consent; and while these resolutions were adopted by the Massachusetts assembly, a circular letter was at the same time sent to all the other colonies, setting forth the need for concerted and harmonious action in respect of so grave a matter. In response, the assemblies of Connecticut, New York, Pennsylvania, Virginia, and South Carolina joined with Massachusetts in remonstrating against the proposed Stamp Act. All these memorials were remarkable for clearness of argument and simple dignity of language. They all took their stand on the principle that, as free-born Englishmen, they could not rightfully be taxed by the House of Commons unless they were represented in that body. But the proviso was added, that if a letter from the secretary of state, coming in the king's name, should be presented to the colonial assemblies, asking them to contribute something from their general resources to the needs of the British Empire, they would cheerfully, as heretofore, grant liberal sums of money, in token of their loyalty and of their interest in all that concerned the welfare of the mighty empire to which they belonged. These able and temperate memorials were sent to England; and in order to reinforce them by personal tact and address, Franklin went over to London as agent for the colony of Pennsylvania.

Reply of the colonies.

The remonstrances of the colonies were of no avail. Early in 1765 the Stamp Act was passed. It is worthy of remark that, now that the obnoxious law was fairly enacted, the idea that the Americans would resist its execution did not at once occur to Franklin. Acquiescence seemed to him, for the present, the only safe policy. In writing to his friend Charles Thomson,

The Stamp Act.

he said that he could no more have hindered the passing of the Stamp Act than he could have hindered the sun's setting. "That," he says, "we could not do. But since it is down, my friend, and it may be long before it rises again, let us make as good a night of it as we can. We may still light candles. Frugality and industry will go a great way towards indemnifying us." But Thomson, in his answer, with truer foresight, observed, "I much fear, instead of the candles you mentioned being lighted, you will hear of the works of darkness!" The news of the passage of the Stamp Act was greeted in America with a burst of indignation. In New York, the act was reprinted with a death's-head upon it in place of the royal arms, and it was hawked about the streets under the title of "The Folly of England and the Ruin of America." In Boston, the church-bells were tolled, and the flags on the shipping put at half-mast.

The Parsons' Cause.

But formal defiance came first from Virginia. A year and a half before, a famous lawsuit, known as the "Parsons' Cause," had brought into public notice a young man who was destined to take rank as one of the greatest of modern orators. The lawsuit which made Patrick Henry's reputation was one of the straws which showed how the stream of tendency in America was then strongly setting toward independence. Tobacco had not yet ceased to be a legal currency in Virginia, and by virtue of an old statute each clergyman of the Established Church was entitled to sixteen thousand pounds of tobacco as his yearly salary. In 1755 and 1758, under the severe pressure of the French war, the assembly had passed relief acts, allowing all public dues, including the salaries of the clergy, to be paid either in kind or in money, at a fixed rate of twopence for a pound of tobacco. The policy of these acts was thoroughly unsound, as they involved a partial repudiation of debts; but the extreme distress of the community was pleaded in excuse, and every one, clergy as well as laymen, at first acquiesced in them. But in 1759, tobacco was worth sixpence per pound, and the clergy became dissatisfied. Their complaints reached the ears of Sherlock, the Bishop of London, and the act of 1758 was summarily vetoed by the king in council. The clergy brought suits to recover the unpaid portions of their salaries; in the test case of Rev. James Maury, the court decided the point of the law in their favour, on the ground of the royal veto, and nothing remained but to settle before a jury the amount of the damages. On this occasion, Henry appeared for the first time in court, and after a few timid and awkward sentences burst forth with an eloquent speech, in which he asserted the indefeasible right of Virginia to make laws for herself, and declared that in annulling a salutary ordinance at the request of a favoured class in the community "a king, from being the father of his people, degenerates into a tyrant, and forfeits all right to obedience." Cries of "Treason!" were heard in the court-room, but the jury immediately returned a verdict of one penny in damages, and Henry became the popular idol of Virginia. The clergy tried in vain to have him indicted for treason, alleging that his crime was hardly less heinous than that which had brought old Lord Lovat to the block. But the people of Louisa county replied, in 1765, by choosing him to represent them in the colonial assembly.

Patrick Henry's resolutions.

Hardly had Henry taken his seat in the assembly when the news of the Stamp Act arrived. In a committee of the whole house, he drew up a series of resolutions, declaring that the colonists were entitled to all the liberties and privileges of natural-born subjects, and that "the taxation of the people by themselves, or by persons chosen by themselves to represent them, . . . is the distinguishing characteristic of British freedom, without which the ancient constitution cannot exist." It was further declared that any attempt to vest the power of taxation in any other body than the colonial assembly was a menace to British no less than to American freedom; that the people of Virginia were not bound to obey any law enacted in disregard of these fundamental principles; and that any one who should maintain the contrary should be regarded as a public enemy. It was in the lively debate which ensued upon these resolutions, that Henry uttered those memorable words commending the example of Tarquin and Cæsar and Charles I. to the attention of George III. Before the vote had been taken upon all the resolutions, Governor Fauquier dissolved the assembly; but the resolutions were printed in the newspapers, and hailed with approval all over the country.

Meanwhile, the Massachusetts legislature, at the suggestion of Otis, had issued a circular letter to all the colonies, calling for a general congress, in order to concert measures of resistance to the Stamp Act. The first cordial response came from South Carolina, at the instance of Christopher Gadsden, a wealthy merchant of Charleston and a scholar learned in Oriental languages, a man of rare sagacity and most liberal spirit. On the 7th of October, the proposed congress assembled at New York, comprising delegates from Massachusetts, South Carolina, Pennsylvania, Rhode Island, Connecticut, Delaware, Maryland, New Jersey, and New York, in all nine colonies, which are here mentioned in the order of the dates at which they chose their delegates. In Virginia, the governor succeeded in preventing the meeting of the legislature, so that this great colony did not send delegates; and for various reasons, New Hampshire, North Carolina, and Georgia were likewise unrepresented at the congress. But the sentiment of all the thirteen colonies was none the less unanimous, and those which did not attend lost no time in declaring their full concurrence with what was done at New York. At this memorable meeting, held under the very guns of the British fleet and hard by the headquarters of General Gage, the commander-in-chief of the regular forces in America, a series of resolutions were adopted, echoing the spirit of Patrick Henry's resolves, though couched in lan-

The Stamp Act Congress.

guage somewhat more conciliatory, and memorials were addressed to the king and to both Houses of Parliament. Of all the delegates present, Gadsden took the broadest ground, in behalf both of liberty and of united action among the colonies. He objected to sending petitions to Parliament, lest thereby its paramount authority should implicitly and unwittingly be acknowledged. "A confirmation of our essential and common rights as Englishmen," said he, "may be pleaded from charters safely enough; but any further dependence on them may be fatal. We should stand upon the broad common ground of those natural rights that we all feel and know as men and as descendants of Englishmen. I wish the charters may not ensnare us at last, by drawing different colonies to act differently in this great cause. Whenever that is the case, all will be over with the whole. There ought to be no New England man, no New Yorker, known on the continent; but all of us Americans." So thought and said this broad-minded South Carolinian.

Declaration of the Massachusetts assembly.

While these things were going on at New York, the Massachusetts assembly, under the lead of Samuel Adams, who had just taken his seat in it, drew up a very able state paper, in which it was declared, among other things, that "the Stamp Act wholly cancels the very conditions upon which our ancestors, with much toil and blood and at their sole expense, settled this country and enlarged his majesty's dominions. It tends to destroy that mutual confidence and affection, as well as that equality, which ought ever to subsist among all his majesty's subjects in this wide and extended empire; and what is the worst of all evils, if his majesty's American subjects are not to be governed according to the known and stated rules of the constitution, their minds may in time become disaffected." This moderate and dignified statement was derided in England as the "raving of a parcel of wild enthusiasts," but from the position here taken Massachusetts never afterward receded.

Resistance to the Stamp Act in Boston;

But it was not only in these formal and decorous proceedings that the spirit of resistance was exhibited. The first announcement of the Stamp Act had called into existence a group of secret societies of workingmen known as "Sons of Liberty," in allusion to a famous phrase in one of Colonel Barré's speeches. These societies were solemnly pledged to resist the execution of the obnoxious law. On the 14th of August, the quiet town of Boston witnessed some extraordinary proceedings. At daybreak, the effigy of the stamp officer, Oliver, was seen hanging from a great elm-tree, while near it was suspended a boot, to represent the late prime minister, Lord Bute; and from the top of the boot-leg there issued a grotesque head, garnished with horns, to represent the devil. At nightfall the Sons of Liberty cut down these figures, and bore them on a bier through the streets until they reached King Street, where they demolished the frame of a house which was supposed to be erecting for a stamp office. Thence, carrying the beams of this frame to Fort Hill, where Oliver lived, they made a bonfire of them in front of his house, and in the bonfire they burned up the effigies. Twelve days after, a mob sacked the splendid house of Chief Justice Hutchinson, threw his plate into the street, and destroyed the valuable library which he had been thirty years in collecting, and which contained many manuscripts, the loss of which was quite irreparable. As usual with mobs, the vengeance fell in the wrong place, for Hutchinson had done his best to prevent the passage of the Stamp Act. In most of the colonies, the stamp officers were compelled to resign their posts. Boxes of stamps arriving by ship were burned or thrown into the sea. Leading merchants agreed to import no more goods from England, and wealthy citizens set the example of dressing in homespun garments. Lawyers agreed to overlook the absence of the stamp on legal documents, while editors derisively issued their newspapers with a death's-head in the place where the stamp was required to be put. In New York, the presence of the troops for a moment encouraged the lieutenant-governor, Colden, to take a bold stand in behalf of the law. He talked of firing upon the people, but was warned that if he did so he would be speedily hanged on a lamp-post, like Captain Porteous of Edinburgh. A torchlight procession, carrying images of Colden and of the devil, broke into the governor's coach-house, and, seizing his best chariot, paraded it about town with the images upon it, and finally burned up chariot and images on the Bowling Green, in full sight of Colden and the garrison, who looked on from the Battery, speechless with

and in New York.

rage, but afraid to interfere. Gage did not dare to have the troops used, for fear of bringing on a civil war; and the next day the discomfited Colden was obliged to surrender all the stamps to the common council of New York, by whom they were at once locked up in the City Hall.

Nothing more was needed to prove the impossibility of carrying the Stamp Act into effect. An act which could be thus rudely defied under the very eyes of the commander-in-chief plainly could never be enforced without a war. But nobody wanted a war, and the matter began to be reconsidered in England. In July, the Grenville ministry had gone out of office, and the Marquis of Rockingham was now prime minister, while Conway, who had been one of the most energetic opponents of the Stamp Act, was secretary of state for the colonies. The new ministry would perhaps have been glad to let the question of taxing America remain in abeyance, but that was no longer possible. The debate on the proposed repeal of the Stamp Act was one of the fiercest that has ever been heard in the House of Commons. Grenville and his friends, now in opposition, maintained in all sincerity that no demand could ever be more just, or more honourably intended, than that which had lately been made upon the Americans. Of the honest conviction of Grenville and his supporters that they were entirely in the right, and that the Americans were governed by purely sordid and vulgar motives in resisting the Stamp Act, there cannot be the slightest doubt. To refute this gross misconception of the American po-

Debate in the House of Commons.

sition, Pitt hastened from a sick-bed to the House of Commons, and delivered those speeches, in which he avowed that he rejoiced in the resistance of the Americans, and declared that, had they submitted tamely to the measures of Grenville, they would have shown themselves only fit to be slaves. He pointed out distinctly that the Americans were upholding those eternal principles of political justice which should be to all Englishmen most dear, and that a victory over the colonies would be of ill-omen for English liberty, whether in the Old World or in the New. Beware, he said, how you persist in this ill-considered policy. "In such a cause your success would be hazardous. America, if she fell, would fall like the strong man with his arms around the pillars of the Constitution." There could be no sounder political philosophy than was contained in these burning sentences of Pitt. From all the history of the European world since the later days of the Roman Republic, there is no more important lesson to be learned than this, — that it is impossible for a free people to govern a dependent people despotically without endangering its own freedom. Pitt therefore urged that the Stamp Act should instantly be repealed, and that the reason for the repeal should be explicitly stated to be because the act "was founded on an erroneous principle." At the same time he recommended the passage of a Declaratory Act, in which the sovereign authority of Parliament over the colonies should be strongly asserted with respect to everything except direct taxation. Similar views were set forth in the

House of Lords, with great learning and ability, by Lord Camden; but he was vehemently opposed by Lord Mansfield, and when the question came to a decision the only peers who supported Camden were Lords Shelburne, Cornwallis, Paulet, and Torrington. The result finally reached was the unconditional repeal of the Stamp Act, and the simultaneous passage of a Declaratory Act, in which the views of Pitt and Camden were ignored, and Parliament asserted its right to make laws binding on the colonies "in all cases whatsoever." By the people of London the repeal was received with enthusiastic delight, and Pitt and Conway, as they appeared on the street, were loudly cheered, while Grenville was greeted with a storm of hisses. In America the effect of the news was electric. There were bonfires in every town, while addresses of thanks to the king were voted in all the legislatures. Little heed was paid to the Declaratory Act, which was regarded merely as an artifice for saving the pride of the British government. There was a unanimous outburst of loyalty all over the country, and never did the people seem less in a mood for rebellion than now.

Repeal of the Stamp Act.

The quarrel had now been made up. On the question of principle, the British had the last word. The government had got out of its dilemma remarkably well, and the plain and obvious course for British statesmanship was not to allow another such direct issue to come up between the colonies and the mother-country. To force on another such issue while the memory of this one was fresh in everybody's mind was sheer madness. To raise the question wantonly, as Charles Townshend did in the course of the very next year, was one of those blunders that are worse than crimes.

The Duke of Grafton's Ministry.

In July, 1766, — less than six months after the repeal of the Stamp Act, — the Rockingham ministry fell, and the formation of a new ministry was entrusted to Pitt, the man who best appreciated the value of the American colonies. But the state of Pitt's health was not such as to warrant his taking upon himself the arduous duties of prime minister. He took the great seal, and, accepting the earldom of Chatham, passed into the House of Lords. The Duke of Grafton became prime minister, under Pitt's guidance; Conway and Lord Shelburne were secretaries of state, and Camden became Lord Chancellor, — all three of them warm friends of America, and adopting the extreme American view of the constitutional questions lately at issue; and along with these was Charles Townshend, the evil spirit of the administration, as chancellor of the exchequer. From such a ministry, it might at first sight seem strange that a fresh quarrel with America should have proceeded. But Chatham's illness soon overpowered him, so that he was kept at home suffering excruciating pain, and could neither guide nor even pay due attention to the proceedings of his colleagues. Of the rest of the ministry, only Conway and Townshend were in the House of Commons, where the real direction of affairs rested; and when Lord Chatham was out of the way, as the Duke of Grafton counted for nothing, the strongest man in the cabinet was unquestionably Townshend. Now when an act for raising an American revenue was proposed by Townshend, a prejudice against it was sure to be excited at once, simply because every American knew well what Townshend's views were. It would have been difficult for such a man even to assume a conciliatory attitude without having his motives suspected; and if the question with Great Britain had been simply that of raising a revenue on statesmanlike principles, it would have been well to entrust the business to some one like Lord Shelburne, in whom the Americans had confidence. In 1767, Townshend ventured to do what in any English ministry of the present day would be impossible. In flat opposition to the policy of Chatham and the rest of his colleagues, trusting in the favour of the king and in his own ability to coax or browbeat the House of Commons, he brought in a series of new measures for taxing America. "I expect to be dismissed for my pains," he said in the House, with flippant defiance; and indeed he came very near it. As soon as he heard what was going on, Chatham mustered up strength enough to go to London and insist upon Townshend's dismissal. But Lord North was the only person that could be thought of to take Townshend's place, and Lord North, who never could bear to offend the king, declined the appointment. Before Chatham could devise a way out of his quandary, his malady again laid him prostrate, and Townshend was not only not turned out, but was left practically supreme in the cabinet. The new measures for taxing America were soon passed. In the debates on the Stamp Act, it had been argued that while Parliament had no right to impose a direct tax upon the Americans, it might still properly regulate American trade by port duties. The distinction had been insisted upon by Pitt, and had been virtually acknowledged by the Americans, who had from time to time submitted to acts of Parliament imposing duties upon merchandise imported into the colonies. Nay, more, when charged with inconsistency for submitting to such acts while resisting the Stamp Act, several leading Americans had explicitly adopted the distinction between internal and external taxation, and declared themselves ready to submit to the latter while determined to resist the former. Townshend was now ready, as he declared, to take them at their word. By way of doing so, he began by laughing to scorn the distinction between internal and external taxation, and declaring that Parliament possessed the undoubted right of taxing the Americans without their own consent; but since objections had been raised to a direct tax, he was willing to resort to port duties, — a measure to which the Americans were logically bound to assent. Duties were accordingly imposed on wine, oil, and fruits, if carried directly to America from Spain or Portugal; on glass, paper, lead, and painters' colours; and lastly on tea. The revenue to be derived from these duties was to be

The Townshend Acts.

devoted to paying a fixed salary to the royal governors and to the justices appointed at the king's pleasure. The Crown was also empowered to create a general civil list in every colony, and to grant salaries and pensions at its arbitrary will. A board of revenue commissioners for the whole country was to be established at Boston, armed with extraordinary powers; and general writs of assistance were expressly legalized and permitted.

Such was the way in which Townshend proceeded to take the Americans at their word. His course was a distinct warning to the Americans that, if they yielded now, they might expect some new Stamp Act or other measures of direct taxation to follow; and so it simply invited resistance. That no doubt might be left on this point, the purpose for which the revenue was to be used showed clearly that the object of this legislation was not to regulate trade, but to assert British supremacy over the colonies at the expense of their political freedom. By providing for a civil list in each colony, to be responsible only to the Crown, it aimed at American self-government even a more deadly blow than had been aimed at it by the Stamp Act. It meddled with the "internal police" of every colony, and would thus have introduced a most vexatious form of tyranny as soon as it had taken effect. A special act by which the Townshend revenue acts were accompanied still further revealed the temper and purposes of the British government. The colony of New York had been required to provide certain

Attack on the New York assembly.

supplies for the regular troops quartered in the city, under command of General Gage; and the colonial assembly had insisted upon providing these supplies in its own way, and in disregard of special instructions from England. For this offence, Parliament now passed an act suspending the New York assembly from its legislative functions until it should have complied with the instructions regarding the supplies to the army. It need not be said that the precedent involved in this act, if once admitted, would have virtually annulled the legislative independence of every one of the colonial assemblies.

Parliament did not properly represent the British people.

We may perhaps wonder that an English Parliament should have been prevailed on to pass such audacious acts as these, and by large majorities. But we must remember that in those days the English system of representation was so imperfect, and had come to be so overgrown with abuses, that an act of Parliament was by no means sure to represent the average judgment of the English people. The House of Commons was so far under the corrupt influence of the aristocracy, and was so inadequately controlled by popular opinion, that at almost any time it was possible for an eloquent, determined, and unscrupulous minister to carry measures through it such as could never have been carried through any of the reformed Parliaments since 1832. It is not easy, perhaps, to say with confidence what the popular feeling in England was in 1767 with reference to the policy of Charles Townshend. The rural population was much more ignorant than it is to-day, and its political opinions were strongly influenced by the country squires, — a worthy set of men, but not generally distinguished for the flexibility of their minds or the breadth of their views. But as a sample of the most intelligent popular feeling in England at that time, it will probably not be unfair to cite that of the city of London, which was usually found arrayed on the side of free government. No wiser advice was heard in Parliament, on the subject of the New York dispute, than was given by alderman Beckford, father of the illustrious author of Vathek, when he said, "Do like the best of physicians, and heal the disease by doing nothing." On many other important occasions in the course of this unfortunate quarrel, the city of London gave expression to opinions which the king and Parliament would have done well to heed. But even if the House of Commons had reflected popular feeling in 1767 as clearly as it has done since 1832, it is by no means sure that it would have known how to deal successfully with the American question. The problem was really a new one in political history; and there was no adequate precedent to guide the statesmen in dealing with the peculiar combination of considerations it involved. As far as concerned the relations of Englishmen in England to the Crown and to Parliament, the British Constitution had at last reached a point where it worked quite smoothly. All contingencies likely to arise seemed to have been provided for. But when it came to the relations of Englishmen in America to the Crown and to Parliament, the case was very different. The case had its peculiar conditions, which the British Constitution in skilful hands would no doubt have proved elastic enough to satisfy; but just at this time the British Constitution happened to be in very unskilful hands, and wholly failed to meet the exigencies of the occasion. The chief difficulty lay in the fact that while on the one hand the American principle of no taxation without representation was unquestionably sound and just, on the other hand the exemption of any part of the British Empire from the jurisdiction of Parliament seemed equivalent to destroying the political unity of the empire. This could not but seem to any English statesman a most lamentable result, and no English statesman felt this more strongly than Lord Chatham.

Difficulty of the problem.

There were only two possible ways in which the difference could be accommodated. Either the American colonies must elect representatives to the Parliament at Westminster; or else the right of levying taxes must be left where it already resided, in their own legislative bodies. The first alternative was seriously considered by eminent political thinkers, both in England and America. In England it was favourably regarded by Adam Smith, and in America by Benjamin Franklin and James Otis. In 1774, some of the loyalists in the first Continental Congress recommended such a scheme. In 1778, after the overthrow of Burgoyne, the king himself began to think favourably of such a way out of the quarrel. But this alternative was doubtless from the first quite visionary and unpractical. The difficulties in the way of securing anything like equality of representation would probably have been insuperable; and the difficulty in dividing jurisdiction fairly between the local colonial legislature and the American contingent in the Parliament at Westminster would far have exceeded any of the difficulties that have arisen in the attempt to adjust the relations of the several States to the general government in our Federal Union. Mere distance, too, which even to-day would go far toward rendering such a scheme impracticable, would have been a still more fatal obstacle in the days of Chatham and Townshend. If, even with the vast enlargement of the political horizon which our hundred years' experience of federalism has effected, the difficulty of such a union still seems so great, we may be sure it would have proved quite insuperable then. The only practicable solution would have been the frank and cordial admission, by the British government, of the essential soundness of the American position, that, in accordance with the entire spirit of the English Constitution, the right of levying taxes in America resided only in the colonial legislatures, in which alone could American freemen be adequately represented. Nor was there really any reason to fear that such a step would imperil the unity of the empire. How mistaken this fear was, on the part of English statesmen, is best shown by the fact that, in her liberal and enlightened dealings with her colonies at the present day, England has consistently adopted the very course of action which alone

Representation of Americans in Parliament.

would have conciliated such men as Samuel Adams in the days of the Stamp Act. By pursuing such a policy, the British government has to-day a genuine hold upon the affections of its pioneers in Australia and New Zealand and Africa. If such a statesman as Gladstone could have dealt freely with the American question during the twelve years following the Peace of Paris, the history of that time need not have been the pitiable story of a blind and obstinate effort to enforce submission to an ill-considered and arbitrary policy on the part of the king and his ministers. The feeling by which the king's party was guided, in the treatment of the American question, was very much the same as the feeling which lately inspired the Tory criticisms upon Gladstone's policy in South Africa. Lord Beaconsfield, a man in some respects not unlike Charles Townshend, bequeathed to his successor a miserable quarrel with the Dutch farmers of the Transvaal; and Mr. Gladstone, after examining the case on its merits, had the moral courage to acknowledge that England was wrong, and to concede the demands of the Boers, even after serious military defeat at their hands. Perhaps no other public act of England in the nineteenth century has done her greater honour than this. But said the Jingoes, All the world will now laugh at Englishmen, and call them cowards. In order to vindicate the military prestige of England, the true policy would be, forsooth, to prolong the war until the Boers had been once thoroughly defeated, and then acknowledge the soundness of their position. Just as if the

Mr. Gladstone and the Boers.

whole world did not know, as well as it can possibly know anything, that whatever qualities the English nation may lack, it certainly does not lack courage, or the ability to win victories in a good cause! All honour to the Christian statesman who dares to leave England's military prestige to be vindicated by the glorious records of a thousand years, and even in the hour of well-merited defeat sets a higher value on political justice than on a reputation for dealing hard blows! Such incidents as this are big with hope for the future. They show us what sort of political morality our children's children may expect to see, when mankind shall have come somewhat nearer toward being truly civilized.

In the eighteenth century, no such exhibition of good sense and good feeling, in the interest of political justice, could have been expected from any European statesman, unless from a Turgot or a Chatham. But Charles Townshend was not even called upon to exercise any such self-control. Had he simply taken Alderman Beckford's advice, and done nothing, all would have been well; but his meddling had now put the government into a position which it was ruinous to maintain, but from which it was difficult to retreat. American tradition rightly lays the chief blame for the troubles which brought on the Revolutionary War to George III.; but, in fairness, it is well to remember that he did not suggest Townshend's measures, though he zealously adopted and cherished them when once propounded. The blame for wantonly throwing the apple of discord belongs to Townshend more than to any one else. After doing this within three months from the time his bill had passed the House of Commons, Townshend was seized with a fever and died at the age of forty-one. A man of extraordinary gifts, but without a trace of earnest moral conviction, he had entered upon a splendid career; but his insincere nature, which turned everything into jest, had stamped itself upon his work. He bequeathed to his country nothing but the quarrel which was soon to deprive her of the grandest part of that empire upon which the sun shall never set.

Death of Townshend.

If Townshend's immediate object in originating these measures was to curry favour with George III., and get the lion's share in the disposal of the king's ample corruption-fund, he had doubtless gone to work in the right way. The king was delighted with Townshend's measures, and after the sudden death of his minister he made them his own, and staked his whole political career as a monarch upon their success. These measures were the fatal legacy which the brighter political charlatan left to the duller political fanatic. The fierce persistency with which George now sought to force Townshend's measures upon the Americans partook of the nature of fanaticism, and we shall not understand it unless we bear in mind the state of political parties in England between 1760 and 1784. When George III. came to the throne, in 1760, England had been governed for more than half a century by the great Whig families which had been brought into the foreground by the revolution of 1688.

His political legacy to George III.

The Tories had been utterly discredited and cast out of political life by reason of their willingness to conspire with the Stuart pretenders in disturbing the peace of the country. Cabinet government, in its modern form, had begun to grow up during the long and prosperous administration of Sir Robert Walpole, who was the first English prime minister in the full sense. Under Walpole's wise and powerful sway, the first two Georges had possessed scarcely more than the shadow of sovereignty. It was the third George's ambition to become a real king, like the king of France or the king of Spain. From earliest babyhood, his mother had forever been impressing upon him the precept, "George, be king!" and this simple lesson had constituted pretty much the whole of his education. Popular tradition regards him as the most ignorant king that ever sat upon the English throne; and so far as general culture is concerned, this opinion is undoubtedly correct. He used to wonder what people could find to admire in such a wretched driveller as Shakespeare, and he never was capable of understanding any problem which required the slightest trace of imagination or of generalizing power. Nevertheless, the popular American tradition undoubtedly errs in exaggerating his stupidity and laying too little stress upon the worst side of his character. George III. was not destitute of a certain kind of ability, which often gets highly rated in this not too clear-sighted world. He could see an immediate end very distinctly, and acquired considerable power from the dogged industry with which he pursued it. In an age where

Character of George III.

some of the noblest English statesmen drank their gallon of strong wine daily, or sat late at the gambling-table, or lived in scarcely hidden concubinage, George III. was decorous in personal habits and pure in domestic relations, and no banker's clerk in London applied himself to the details of business more industriously than he. He had a genuine talent for administration, and he devoted this talent most assiduously to selfish ends. Scantily endowed with human sympathy, and almost boorishly stiff in his ordinary unstudied manner, he could be smooth as oil whenever he liked. He was an adept in gaining men's confidence by a show of interest, and securing their aid by dint of fair promises; and when he found them of no further use, he could turn them adrift with wanton insult. Any one who dared to disagree with him upon even the slightest point of policy he straightway regarded as a natural enemy, and pursued him ever afterward with vindictive hatred. As a natural consequence, he surrounded himself with weak and short-sighted advisers, and toward all statesmen of broad views and independent character he nursed the bitterest rancour. He had little faith in human honour or rectitude, and in pursuing an end he was seldom deterred by scruples.

Such was the man who, on coming to the throne in 1760, had it for his first and chiefest thought to break down the growing system of cabinet government in England. For the moment circumstances seemed to favour him. The ascendancy of the great Whig families was endangered on two sides. On the one hand, the Tory party had outlived that idle, romantic love for the Stuarts upon which it found it impossible to thrive. The Tories began coming to court again, and they gave the new king all the benefit of their superstitious theories of high prerogative and divine right. On the other hand, a strong popular feeling was beginning to grow up against parliamentary government as conducted by the old Whig families. The House of Commons no longer fairly represented the people. Ancient boroughs, which possessed but a handful of population, or, like Old Sarum, had no inhabitants at all, still sent their representatives to Parliament, while great cities of recent growth, such as Birmingham and Leeds, were unrepresented. To a great extent, it was the most progressive parts of the kingdom which were thus excluded from a share in the government, while the rotten boroughs were disposed of by secret lobbying, or even by open bargain and sale. A few Whig families, the heads of which sat in the House of Lords, thus virtually owned a considerable part of the House of Commons; and, under such circumstances, it was not at all strange that Parliament should sometimes, as in the Wilkes case, array itself in flat opposition to the will of the people. The only wonder is that there were not more such scandals. The party of "Old Whigs," numbering in its ranks some of the ablest and most patriotic men in England, was contented with this state of things, upon which it had thrived for two generations, and could not be made to understand the iniquity of it, — any more than an old cut-and-dried American

English parties between 1760 and 1784.

politician in our time can be made to understand the iniquity of the "spoils system." Of this party the Marquis of Rockingham was the political leader, and Edmund Burke was the great representative statesman. In strong opposition to the Old Whig policy there had grown up the party of New Whigs, bent upon bringing about some measure of parliamentary reform, whereby the House of Commons might truly represent the people of Great Britain. In Parliament this party was small in numbers, but weighty in character, and at its head was the greatest Englishman of the eighteenth century, the elder William Pitt, under whose guidance England had won her Indian empire and established her dominion over the seas, while she had driven the French from America, and enabled Frederick the Great to lay the foundations of modern Germany.

George III. as a politician.

Now when George III. came to the throne, he took advantage of this division in the two parties in order to break down the power of the Old Whig families, which so long had ruled the country. To this end he used the revived Tory party with great effect, and bid against the Old Whigs for the rotten boroughs; and in playing off one set of prejudices and interests against another, he displayed in the highest degree the cunning and craft of a self-seeking politician. His ordinary methods would have aroused the envy of Tammany. While engaged in such work, he had sense enough to see that the party from which he had most to fear was that of the New Whigs, whose scheme of parliamentary reform, if ever successful, would deprive him of the machinery of corruption upon which he relied. Much as he hated the Old Whig families, he hated Pitt and his followers still more heartily. He was perpetually denouncing Pitt as a "trumpeter of sedition," and often vehemently declared in public, and in the most offensive manner, that he wished that great man were dead. Such had been his eagerness to cast discredit upon Pitt's policy that he had utterly lost sight of the imperial interests of England, which indeed his narrow intelligence was incapable of comprehending. One of the first acts of his reign had been to throw away Cuba and the Philippine Islands, which Pitt had just conquered from Spain; while at the same time, by leaving Prussia in the lurch before the Seven Years' War had fairly closed, he converted the great Frederick from one of England's warmest friends into one of her bitterest enemies.

His chief reason for quarreling with the Americans.

This political attitude of George III. toward the Whigs in general, and toward Pitt in particular, explains the fierce obstinacy with which he took up and carried on Townshend's quarrel with the American colonies. For if the American position, that there should be no taxation without representation, were once to be granted, then it would straightway become necessary to admit the principles of parliamentary reform. The same principle that applied to such commonwealths as Massachusetts and Virginia would be forthwith applied to such towns as Birmingham and Leeds. The system of rotten boroughs would be swept away; the chief engine of kingly corruption would thus be destroyed; a reformed House of Commons, with the people at its back, would curb forever the pretensions of the Crown; and the detested Lord Chatham would become the real ruler of a renovated England, in which George III. would be a personage of very little political importance.

In these considerations we find the explanation of the acts of George III. which brought on the American Revolution, and we see why it is historically correct to regard him as the person chiefly responsible for the quarrel. The obstinacy with which he refused to listen to a word of reason from America was largely due to the exigencies of the political situation in which he found himself. For him, as well as for the colonies, it was a desperate struggle for political existence. He was glad to force on the issue in America rather than in England, because it would be comparatively easy to enlist British local feeling against the Americans as a remote set of "rebels," with whom Englishmen had no interests in common, and thus obscure the real nature of the issue. Herein he showed himself a cunning politician, though an ignoble statesman. By playing off against each other the two sections of the Whig party, he continued for a while to carry his point; and had he succeeded in overcoming the American resistance and calling into England a well-trained army of victorious mercenaries, the political quarrel there could hardly have failed to develop into a civil war. A new rebellion would perhaps have overthrown George III. as James II. had been overthrown a century before. As it was, the victory of the Americans put an end to the personal government of the king in 1784, so quietly that the people scarcely realized the change.[1] A peaceful election accomplished what otherwise could hardly have been effected without bloodshed. So while George III. lost the fairest portion of the British Empire, it was the sturdy Americans who, fighting the battle of freedom at once for the Old World and for the New, ended by overwhelming his paltry schemes for personal aggrandizement in hopeless ruin, leaving him for posterity to contemplate as one of the most instructive examples of short-sighted folly that modern history affords.

[1] See my *Critical Period of American History*, chap. i.

CHAPTER II.

THE CRISIS.

Lord North.

TOWNSHEND was succeeded in the exchequer by Lord North, eldest son of the Earl of Guildford, a young man of sound judgment, wide knowledge, and rare sweetness of temper, but wholly lacking in sympathy with popular government. As leader of the House of Commons, he was sufficiently able in debate to hold his ground against the fiercest attacks of Burke and Fox, but he had no strength of will. His lazy good-nature and his Tory principles made him a great favourite with the king, who, through his influence over Lord North, began now to exercise the power of a cabinet minister, and to take a more important part than hitherto in the direction of affairs. Soon after North entered the cabinet, colonial affairs were taken from Lord Shelburne and put in charge of Lord Hillsborough, a man after the king's own heart. Conway was dismissed from the cabinet, and his place was taken by Lord Weymouth, who had voted against the repeal of the Stamp Act. The Earl of Sandwich, who never spoke of the Americans but in terms of abuse, was at the same time made postmaster-general; and in the following year Lord Chatham resigned the privy seal.

While the ministry, by these important changes, was becoming more and more hostile to the just claims of the Americans, those claims were powerfully urged in America, both in popular literature and in well-considered state papers. John Dickinson, at once a devoted friend of England and an ardent American patriot, published his celebrated Farmer's Letters, which were greatly admired in both countries for their temperateness of tone and elegance of expression. In these letters, Dickinson held a position quite similar to that occupied by Burke. Recognizing that the constitutional relations of the colonies to the mother country had always been extremely vague and ill-defined, he urged that the same state of things should be kept up forever through a genuine English feeling of compromise, which should refrain from pushing any abstract theory of sovereignty to its extreme logical conclusions. At the same time, he declared that the Townshend revenue acts were "a most dangerous innovation" upon the liberties of the people, and significantly hinted that, should the ministry persevere in its tyrannical policy, "English history affords examples of resistance by force."

John Dickinson.

While Dickinson was publishing these letters, Samuel Adams wrote for the Massachusetts assembly a series of addresses to the ministry, a petition to the king, and a circular letter to the assemblies of the other colonies. In these very able state papers, Adams declared that a proper representation of American interests in the British Parliament was imprac-

The Massachusetts circular letter.

ticable, and that, in accordance with the spirit of the English Constitution, no taxes could be levied in America except by the colonial legislatures. He argued that the Townshend acts were unconstitutional, and asked that they should be repealed, and that the colonies should resume the position which they had occupied before the beginning of the present troubles. The petition to the king was couched in beautiful and touching language, but the author seems to have understood very well how little effect it was likely to produce. His daughter, Mrs. Wells, used to tell how one evening, as her father had just finished writing this petition, and had taken up his hat to go out, she observed that the paper would soon be touched by the royal hand. "More likely, my dear," he replied, "it will be spurned by the royal foot!" Adams rightly expected much more from the circular letter to the other colonies, in which he invited them to coöperate with Massachusetts in resisting the Townshend acts, and in petitioning for their repeal. The assembly, having adopted all these papers by a large majority, was forthwith prorogued by Governor Bernard, who, in a violent speech, called them demagogues to whose happiness "everlasting contention was necessary." But the work was done. The circular letter brought encouraging replies from the other colonies. The condemnation of the Townshend acts was unanimous, and leading merchants in most of the towns entered into agreements not to import any more English goods until the acts should be repealed. Ladies formed associations, under the name of Daughters of Liberty, pledging themselves to wear homespun clothes and to abstain from drinking tea. The feeling of the country was thus plainly enough expressed, but nowhere as yet was there any riot or disorder, and no one as yet, except, perhaps, Samuel Adams, had begun to think of a political separation from England. Even he did not look upon such a course as desirable, but the treatment of his remonstrances by the king and the ministry soon led him to change his opinion.

The petition of the Massachusetts assembly was received by the king with silent contempt, but the circular letter threw him into a rage. In cabinet meeting, it was pronounced to be little better than an overt act of rebellion, and the ministers were encouraged in this opinion by letters from Bernard, who represented the whole affair as the wicked attempt of a few vile demagogues to sow the seeds of dissension broadcast over the continent. We have before had occasion to observe the extreme jealousy with which the Crown had always regarded any attempt at concerted action among the colonies which did not originate with itself. But here was an attempt at concerted action in flagrant opposition to the royal will. Lord Hillsborough instructed Bernard to command the assembly to rescind their circular letter, and, in case of their refusal, to send them home about their business. This was to be repeated year after year, so that, until Massachusetts should see fit to declare herself humbled and penitent, she must go without

Lord Hillsborough's instructions to Bernard.

a legislature. At the same time, Hillsborough ordered the assemblies in all the other colonies to treat the Massachusetts circular with contempt, — and this, too, under penalty of instant dissolution. From a constitutional point of view, these arrogant orders deserve to be ranked among the curiosities of political history. They serve to mark the rapid progress the ministry was making in the art of misgovernment. A year before, Townshend had suspended the New York legislature by an act of Parliament. Now, a secretary of state, by a simple royal order, threatened to suspend all the legislative bodies of America unless they should vote according to his dictation.

The "Illustrious Ninety-Two."

When Hillsborough's orders were laid before the Massachusetts assembly, they were greeted with scorn. "We are asked to rescind," said Otis. "Let Britain rescind her measures, or the colonies are lost to her forever." Nevertheless, it was only after nine days of discussion that the question was put, when the assembly decided, by a vote of ninety-two to seventeen, that it would not rescind its circular letter. Bernard immediately dissolved the assembly, but its vote was hailed with delight throughout the country, and the "Illustrious Ninety-Two" became the favourite toast on all convivial occasions. Nor were the other colonial assemblies at all readier than that of Massachusetts to yield to the secretary's dictation. They all expressed the most cordial sympathy with the recommendations of the circular letter; and in several instances they were dissolved by the governors, according to Hillsborough's instructions.

While these fruitless remonstrances against the Townshend acts had been preparing, the commissioners of the customs, in enforcing the acts, had not taken sufficient pains to avoid irritating the people. In the spring of 1768, the fifty-gun frigate Romney had been sent to mount guard in the harbour of Boston, and while she lay there several of the citizens were seized and impressed as seamen,—a lawless practice long afterward common in the British navy, but already stigmatized as barbarous by public opinion in America. As long ago as 1747, when the relations between the colonies and the home government were quite harmonious, resistance to the press-gang had resulted in a riot in the streets of Boston. Now while the town was very indignant over this lawless kidnapping of its citizens, on the 10th of June, 1768, John Hancock's sloop Liberty was seized at the wharf by a boat's crew from the Romney, for an alleged violation of the revenue laws, though without official warrant. Insults and recriminations ensued between the officers and the citizens assembled on the wharf, until after a while the excitement grew into a mild form of riot, in which a few windows were broken, some of the officers were pelted, and finally a pleasure-boat, belonging to the collector, was pulled up out of the water, carried to the Common, and burned there, when Hancock and Adams, arriving upon the scene, put a stop to the commotion. A few days afterward, a town meeting was held in Faneuil Hall; but as the crowd was too great to be contained in the building, it was adjourned to the Old South

Impressment of citizens.

Meeting-House, where Otis addressed the people from the pulpit. A petition to the governor was prepared, in which it was set forth that the impressment of peaceful citizens was an illegal act, and that the state of the town was as if war had been declared against it; and the governor was requested to order the instant removal of the frigate from the harbour. A committee of twenty-one leading citizens was appointed to deliver this petition to the governor at his house in Jamaica Plain. In his letters to the secretary of state, Bernard professed to live in constant fear of assassination, and was always begging for troops to protect him against the incendiary and blackguard mob of Boston. Yet as he looked down the beautiful road from his open window, that summer afternoon, what he saw was not a ragged mob, armed with knives and bludgeons, shouting "Liberty, or death!" and bearing the head of a revenue collector aloft on the point of a pike, but a quiet procession of eleven chaises, from which there alighted at his door twenty-one gentlemen, as sedate and stately in demeanour as those old Roman senators at whom the Gaulish chief so marveled. There followed a very affable interview, during which wine was passed around. The next day the governor's answer was read in town meeting, declining to remove the frigate, but promising that in future there should be no impressment of Massachusetts citizens; and with this compromise the wrath of the people was for a moment assuaged.

Affairs of this sort, reported with gross exaggeration by the governor and revenue commissioners to the ministry, produced in England the impression that Boston was a lawless and riotous town, full of cutthroats and blacklegs, whose violence could be held in check only by martial law. Of all the misconceptions of America by England which brought about the American Revolution, perhaps this notion of the turbulence of Boston was the most ludicrous. During the ten years of excitement which preceded the War of Independence there was one disgraceful riot in Boston, — that in which Hutchinson's house was sacked; but in all this time not a drop of blood was shed by the people, nor was anybody's life for a moment in danger at their hands. The episode of the sloop Liberty, as here described, was a fair sample of the disorders which occurred at Boston at periods of extreme excitement; and in any European town in the eighteenth century it would hardly have been deemed worthy of mention.

Even before the affair of the Liberty, the government had made up its mind to send troops to Boston, in order to overawe the popular party and show them that the king and Lord Hillsborough were in earnest. The news of the Liberty affair, however, served to remove any hesitation that might hitherto have been felt. Vengeance was denounced against the insolent town of Boston. The most seditious spirits, such as Otis and Adams, must be made an example of, and thus the others might be frightened into submission. With such intent, Lord Hillsborough sent over to inquire "if any person

Statute of Henry VIII. concerning "treason committed abroad."

had committed any acts which, under the statutes of Henry VIII. against treason committed abroad, might justify their being brought to England for trial." This raking-up of an obsolete statute, enacted at one of the worst periods of English history, and before England had any colonies at all, was extremely injudicious. But besides all this, continued Hillsborough, the town meeting, that nursery of sedition, must be put down or overawed; and in pursuance of this scheme, two regiments of soldiers and a frigate were to be sent over to Boston at the ministry's earliest convenience. To make an example of Boston, it was thought, would have a wholesome effect upon the temper of the Americans.

Samuel Adams makes up his mind, 1768.

It was now, in the summer of 1768, that Samuel Adams made up his mind that there was no hope of redress from the British government, and that the only remedy was to be found in the assertion of political independence by the American colonies. The courteous petitions and temperate remonstrances of the American assemblies had been met, not by rational arguments, but by insulting and illegal royal orders; and now at last an army was on the way from England to enforce the tyrannical measures of government, and to terrify the people into submission. Accordingly, Adams came to the conclusion that the only proper course for the colonies was to declare themselves independent of Great Britain, to unite together in a permanent confederation, and to invite European alliances. We have his own word for the fact that from this moment until the Declaration of Independence, in 1776, he consecrated all his energies, with burning enthusiasm, upon the attainment of that great object. Yet in 1768 no one knew better than Samuel Adams that the time had not yet come when his bold policy could be safely adopted, and that any premature attempt at armed resistance on the part of Massachusetts might prove fatal. At this time probably no other American statesman had thought the matter out so far as to reach Adams's conclusions. No American had as yet felt any desire to terminate the political connection with England. Even those who most thoroughly condemned the measures of the government did not consider the case hopeless, but believed that in one way or another a peaceful solution was still attainable. For a long time this attitude was sincerely and patiently maintained. Even Washington, when he came to take command of the army at Cambridge, after the battle of Bunker Hill, had not made up his mind that the object of the war was to be the independence of the colonies. In the same month of July, 1775, Jefferson said expressly, "We have not raised armies with designs of separating from Great Britain and establishing independent states. Necessity has not yet driven us into that desperate measure." The Declaration of Independence was at last brought about only with difficulty and after prolonged discussion. Our great-great-grandfathers looked upon themselves as Englishmen, and felt proud of their connection with England. Their determination to resist arbitrary measures was at first in no way associated in their minds with disaffection toward the mother-country. Besides this, the task of effecting a separation by military measures seemed to most persons quite hopeless. It was not until after Bunker Hill had shown that American soldiers were a match for British soldiers in the field, and after Washington's capture of Boston had shown that the enemy really could be dislodged from a whole section of the country, that the more hopeful patriots began to feel confident of the ultimate success of a war for independence. It is hard for us now to realize how terrible the difficulties seemed to the men who surmounted them. Throughout the war, beside the Tories who openly sympathized with the enemy, there were many worthy people who thought we were "going too far," and who magnified our losses and depreciated our gains, — quite like the people who, in the War of Secession, used to be called "croakers." The depression of even the boldest, after such defeats as that of Long Island, was dreadful. How inadequate was the general sense of our real strength, how dim the general comprehension of the great events that were happening, may best be seen in the satirical writings of some of the loyalists. At the time of the French alliance, there were many who predicted that the result of this step would be to undo the work of the Seven Years' War, to reinstate the French in America with full control over the thirteen colonies, and to establish despotism and popery all over the continent. A satirical pamphlet, published in 1779, just ten years before the Bastille was torn down in Paris, drew an imaginary picture of a Bastille which ten years later was to stand in New York, and, with still further license of fantasy, portrayed Samuel Adams in the garb of a Dominican friar. Such nonsense is of course no index to the sentiments or the beliefs of the patriotic American people, but the mere fact that it could occur to anybody shows how hard it was for people to realize how competent America was to take care of herself. The more we reflect upon the slowness with which the country came to the full consciousness of its power and importance, the more fully we bring ourselves to realize how unwilling America was to tear herself asunder from England, and how the Declaration of Independence was only at last resorted to when it had become evident that no other course was compatible with the preservation of our self-respect; the more thoroughly we realize all this, the nearer we shall come toward duly estimating the fact that in 1768, seven years before the battle of Lexington, the master mind of Samuel Adams had fully grasped the conception of a confederation of American states independent of British control. The clearness with which he saw this, as the inevitable outcome of the political conditions of the time, gave to his views and his acts, in every emergency that arose, a commanding influence throughout the land that was simply incalculable.

In September, 1768, it was announced in Boston that the troops were on their way, and would soon be landed. There happened to be a legal obstacle, unforeseen by the ministry, to their being quartered in the town. In accordance with the general act of Parliament for quartering troops, the regular barracks at Castle William in the harbour would have to be filled before the town could be required to find quarters for any troops. Another clause of the act provided that if any military officer should take upon himself to quarter soldiers in any of his Majesty's dominions otherwise than as allowed by the act, he should be straightway dismissed the service. At the news that the troops were about to arrive, the governor was asked to convene the assembly, that it might be decided how to receive them. On Bernard's refusal, the selectmen of Boston issued a circular, inviting all the towns of Massachusetts to send delegates to a general convention, in order that deliberate action might be taken upon this important matter. In answer to the circular, delegates from ninety-six towns assembled in Faneuil Hall, and, laughing at the governor's order to "disperse," proceeded to show how, in the exercise of the undoubted right of public meeting, the colony could virtually legislate for itself, in the absence of its regular legislature. The convention, finding that nothing was necessary for Boston to do but insist upon strict compliance with the letter of the law, adjourned. In October, two regiments arrived, and were allowed to land without opposition, but no lodging was provided for them. Bernard, in fear of an affray, had gone out into the country; but nothing could have been farther from the thoughts of the people. The commander, Colonel Dalrymple, requested shelter for his men, but was told that he must quarter them in the barracks at Castle William. As the night was frosty, how-

Arrival of troops in Boston.

ever, the Sons of Liberty allowed them to sleep in Faneuil Hall. Next day, the governor, finding everything quiet, came back, and heard Dalrymple's complaint. But in vain did he apply in turn to the council, to the selectmen, and to the justices of the peace, to grant quarters for the troops; he was told that the law was plain, and that the Castle must first be occupied. The governor then tried to get possession of an old dilapidated building which belonged to the colony; but the tenants had taken legal advice, and told him to turn them out if he dared. Nothing could be more provoking. General Gage was obliged to come on from his headquarters at New York; but not even he, the commander-in-chief of his Majesty's forces in America, could quarter the troops in violation of the statute without running the risk of being cashiered, on conviction before two justices of the peace. So the soldiers stayed at night in tents on the Common, until the weather grew so cold that Dalrymple was obliged to hire some buildings for them at exorbitant rates, and at the expense of the Crown. By way of insult to the people, two cannon were planted on King Street, with their muzzles pointing toward the Town House. But as the troops could do nothing without a requisition from a civil magistrate, and as the usual strict decorum was preserved throughout the town, there was nothing in the world for them to do. In case of an insurrection, the force was too small to be of any use; and so far as the policy of overawing the town was concerned, no doubt the soldiers were more afraid of the people than the people of the soldiers.

No sooner were the soldiers thus established in Boston than Samuel Adams published a series of letters signed "Vindex," in which he argued that to keep up "a standing army within the kingdom in time of peace, without the consent of Parliament, was against the law; that the consent of Parliament necessarily implied the consent of the people, who were always present in Parliament, either by themselves or by their representatives; and that the Americans, as they were not and could not be represented in Parliament, were therefore suffering under military tyranny over which they were allowed to exercise no control." The only notice taken of this argument by Bernard and Hillsborough was an attempt to collect evidence upon the strength of which its author might be indicted for treason, and sent over to London to be tried; but Adams had been so wary in all his proceedings that it was impossible to charge him with any technical offence, and to have seized him otherwise than by due process of law would have been to precipitate rebellion in Massachusetts.

Letters of "Vindex."

In Parliament, the proposal to extend the act of Henry VIII. to America was bitterly opposed by Burke, Barré, Pownall, and Dowdeswell, and even by Grenville, who characterized it as sheer madness; but the measure was carried, nevertheless. Burke further maintained, in an eloquent speech, that the royal order requiring Massachusetts to rescind her circular letter was unconstitutional; and here again Grenville agreed with him. The attention of Parliament, during the

Debate in Parliament.

spring of 1769, was occupied chiefly with American affairs. Pownall moved that the Townshend acts should be repealed, and in this he was earnestly seconded by a petition of the London merchants; for the non-importation policy of Americans had begun to bear hard upon business in London. After much debate, Lord North proposed a compromise, repealing all the Townshend acts except that which laid duty on tea. The more clearheaded members saw that such a compromise, which yielded nothing in the matter of principle, would do no good. Beckford pointed out the fact that the tea-duty did not bring in £300 to the government; and Lord Beauchamp pertinently asked whether it were worth while, for such a paltry revenue, to make enemies of three millions of people. Grafton, Camden, Conway, Burke, Barré, and Dowdeswell wished to have the tea-duty repealed also, and the whole principle of parliamentary taxation given up; and Lord North agreed with them in his secret heart, but could not bring himself to act contrary to the king's wishes. "America must fear you before she can love you," said Lord North. . . . "I am against repealing the last act of Parliament, securing to us a revenue out of America; I will never think of repealing it until I see America prostrate at my feet." "To effect this," said Barré, "is not so easy as some imagine; the Americans are a numerous, a respectable, a hardy, a free people. But were it ever so easy, does any friend to his country really wish to see America thus humbled? In such a situation, she would serve only as a monument of your arrogance and

Colonel Barré's speech.

your folly. For my part, the America I wish to see is America increasing and prosperous, raising her head in graceful dignity, with freedom and firmness asserting her rights at your bar, vindicating her liberties, pleading her services, and conscious of her merit. This is the America that will have spirit to fight your battles, to sustain you when hard pushed by some prevailing foe, and by her industry will be able to consume your manufactures, support your trade, and pour wealth and splendour into your towns and cities. If we do not change our conduct towards her, America will be torn from our side. . . . Unless you repeal this law, you run the risk of losing America." But the ministers were deaf to Barré's sweet reasonableness. "We shall grant nothing to the Americans," said Lord Hillsborough, "except what they may ask with a halter round their necks." "They are a race of convicted felons," echoed poor old Dr. Johnson, — who had probably been reading Moll Flanders, — "and they ought to be thankful for anything we allow them short of hanging."

As the result of the discussion, Lord North's so-called compromise was adopted, and a circular was sent to America, promising that all the obnoxious acts, except the tea-duty, should be repealed. At the same time, Bernard was recalled from Massachusetts to appease the indignation of the people, and made a baronet to show that the ministry approved of his conduct as governor. His place was filled by the lieutenant-governor, Thomas Hutchinson a man of great learning and brilliant talent, whose "History of

Thomas Hutchinson.

Massachusetts Bay" entitles him to a high rank among the worthies of early American literature. The next year Hutchinson was appointed governor. As a native of Massachusetts, it was supposed by Lord North that he would be less likely to irritate the people than his somewhat arrogant predecessor. But in this the government turned out to be mistaken. As to Hutchinson's sincere patriotism there can now be no doubt whatever. There was something pathetic in the intensity of his love for New England, which to him was the goodliest of all lands, the paradise of this world. He had been greatly admired for his learning and accomplishments, and the people of Massachusetts had elected him to one office after another, and shown him every mark of esteem until the evil days of the Stamp Act. It then appeared that he was a Tory on principle, and a thorough believer in the British doctrine of the absolute supremacy of Parliament, and popular feeling instantly turned against him. He was called a turncoat and traitor, and a thankless dog withal, whose ruling passion was avarice. His conduct and his motives were alike misjudged. He had tried to dissuade the Grenville ministry from passing the Stamp Act; but when once the obnoxious measure had become law, he thought it his duty to enforce it like other laws. For this he was charged with being recreant to his own convictions, and in the shameful riot of August, 1765, he was the worst sufferer. No public man in America has ever been the object of more virulent hatred. None has been more grossly misrepresented by historians. His appointment as governor, however well meant, turned out to be anything but a wise measure.

While these things were going on, a strong word of sympathy came from Virginia. When Hillsborough made up his mind to browbeat Boston, he thought it worth while to cajole the Virginians, and try to win them from the cause which Massachusetts was so boldly defending. So Lord Botetourt, a most genial and conciliatory man, was sent over to be governor of Virginia, to beguile the people with his affable manner and sweet discourse. But between a quarrelsome Bernard and a gracious Botetourt the practical difference was little, where grave questions of constitutional right were involved. In May, 1769, the Virginia legislature assembled at Williamsburgh. Among its members were Patrick Henry, Washington, and Jefferson. The assembly condemned the Townshend acts, asserted that the people of Virginia could be taxed only by their own representatives, declared that it was both lawful and expedient for all the colonies to join in a protest against any violation of the rights of Americans, and especially warned the king of the dangers that might ensue if any American citizen were to be carried beyond sea for trial. Finally, it sent copies of these resolutions to all the other colonial assemblies, inviting their concurrence. At this point Lord Botetourt dissolved the assembly; but the members straightway met again in convention at the famous Apollo room of the Raleigh tavern, and adopted a series of resolutions prepared by Washington, in which they pledged themselves to

Virginia resolutions, 1769.

continue the policy of non-importation until all the obnoxious acts of 1767 should be repealed. These resolutions were adopted by all the southern colonies.

All through the year 1769, the British troops remained quartered in Boston at the king's expense. According to Samuel Adams, their principal employment seemed to be to parade in the streets, and by their merry-andrew tricks to excite the contempt of women and children. But the soldiers did much to annoy the people, to whom their very presence was an insult. They led brawling, riotous lives, and made the quiet streets hideous by night with their drunken shouts. Scores of loose women, who had followed the regiments across the ocean, came to scandalize the town for a while, and then to encumber the almshouse. On Sundays the soldiers would race horses on the Common, or play Yankee Doodle just outside the church-doors during the services. Now and then oaths, or fisticuffs, or blows with sticks, were exchanged between soldiers and citizens, and once or twice a more serious affair occurred. One evening in September, a dastardly assault was made upon James Otis, in the British Coffee House, by one Robinson, a commissioner of customs, assisted by half a dozen army officers. It reminds one of the assault upon Charles Sumner by Brooks of South Carolina, shortly before the War of Secession. Otis was savagely beaten, and received a blow on the head with a sword, from the effects of which he never recovered, but finally lost his reason. The popular

Assault on James Otis.

wrath at this outrage was intense, but there was no disturbance. Otis brought suit against Robinson, and recovered £2,000 in damages, but refused to accept a penny of it when Robinson confessed himself in the wrong, and humbly asked pardon for his irreparable offence.

On the 22d of February, 1770, an informer named Richardson, being pelted by a party of school-boys, withdrew into his house, opened a window, and fired at random into the crowd, killing one little boy and severely wounding another. He was found guilty of murder, but was pardoned. At last, on the 2d of March, an angry quarrel occurred between a party of soldiers and some of the workmen at a ropewalk, and for two or three days there was considerable excitement in the town, and people talked together, standing about the streets in groups; but Hutchinson did not even take the precaution of ordering the soldiers to be kept within their barracks, for he did not believe that the people intended a riot, nor that the troops would dare to fire on the citizens without express permission from himself. On the evening of March 5th, at about eight o'clock, a large crowd collected near the barracks, on Brattle Street, and from bandying abusive epithets with the soldiers began pelting them with snow-balls and striking at them with sticks, while the soldiers now and then dealt blows with their muskets. Presently Captain Goldfinch, coming along, ordered the men into their barracks for the night, and thus stopped the affray. But meanwhile some one had got into the Old Brick

The "Boston Massacre."

Meeting-House, opposite the head of King Street, and rung the bell; and this, being interpreted as an alarm of fire, brought out many people into the moonlit streets. It was now a little past nine. The sentinel who was pacing in front of the Custom House had a few minutes before knocked down a barber's boy for calling names at the captain, as he went up to stop the affray on Brattle Street. The crowd in King Street now began to pelt the sentinel, and some shouted, "Kill him!" when Captain Preston and seven privates from the twenty-ninth regiment crossed the street to his aid: and thus the file of nine soldiers confronted an angry crowd of fifty or sixty unarmed men, who pressed up to the very muzzles of their guns, threw snow at their faces, and dared them to fire. All at once, but quite unexpectedly and probably without orders from Preston, seven of the levelled pieces were discharged, instantly killing four men and wounding seven others, of whom two afterwards died. Immediately the alarm was spread through the town, and it might have gone hard with the soldiery, had not Hutchinson presently arrived on the scene, and quieted the people by ordering the arrest of Preston and his men. Next morning the council advised the removal of one of the regiments, but in the afternoon an immense town meeting, called at Faneuil Hall, adjourned to the Old South Meeting-House; and as they passed by the Town House, the lieutenant-governor, looking out upon their march, judged "their spirit to be as high as was the spirit of their ancestors when they imprisoned Andros, while they were four times as numerous." All the way from the church to the Town House the street was crowded with the people, while a committee, headed by Samuel Adams, waited upon the governor, and received his assurance that one regiment should be removed. As the committee came out from the Town House, to carry the governor's reply to the meeting in the church, the people pressed back on either side to let them pass; and Adams, leading the way with uncovered head through the lane thus formed, and bowing first to one side and then to the other, passed along the watchword, "Both regiments, or none!" When, in the church, the question was put to vote, three thousand voices shouted, "Both regiments, or none!" and armed with this ultimatum the committee returned to the Town House, where the governor was seated with Colonel Dalrymple and the members of the council. Then Adams, in quiet but earnest tones, stretching forth his arm and pointing his finger at Hutchinson, said that if as acting governor of the province he had the power to remove one regiment he had equally the power to remove both, that the voice of three thousand freemen demanded that all soldiery be forthwith removed from the town, and that if he failed to heed their just demand, he did so at his peril. "I observed his knees to tremble," said the old hero afterward, "I saw his face grow pale, — and I enjoyed the sight!" Before sundown the order had gone forth for the removal of both regiments to Castle William, and not until then did the meeting in the church break up. From that day forth the fourteenth and twenty-ninth regiments were known in Parliament as "the Sam Adams regiments."

Such was the famous Boston Massacre. All the mildness of New England civilization is brought most strikingly before us in that truculent phrase. The careless shooting of half a dozen townsmen is described by a word which historians apply to such events as Cawnpore or the Sicilian Vespers. Lord Sherbrooke, better known as Robert Lowe, declared a few years ago, in a speech on the uses of a classical education, that the battle of Marathon was really of less account than a modern colliery explosion, because only one hundred and ninety-two of the Greek army lost their lives! From such a point of view, one might argue that the Boston Massacre was an event of far less importance than an ordinary free fight among Colorado gamblers. It is needless to say that this is not the historical point of view. Historical events are not to be measured with a foot-rule. This story of the Boston Massacre is a very trite one, but it has its lessons. It furnishes an instructive illustration of the high state of civilization reached by the people among whom it happened, — by the oppressors as well as those whom it was sought to oppress. The quartering of troops in a peaceful town is something that has in most ages been regarded with horror. Under the senatorial government of Rome, it used to be said that the quartering of troops, even upon a friendly province and for the purpose of protecting it, was a visitation only less to be dreaded than an inroad of hostile barbarians. When we

Some lessons of the "Massacre."

reflect that the British regiments were encamped in Boston during seventeen months, among a population to whom they were thoroughly odious, the fact that only half a dozen persons lost their lives, while otherwise no really grave crimes seem to have been committed, is a fact quite as creditable to the discipline of the soldiers as to the moderation of the people. In most ages and countries, the shooting of half a dozen citizens under such circumstances would either have produced but a slight impression, or, on the other hand, would perhaps have resulted on the spot in a wholesale slaughter of the offending soldiers. The fact that so profound an impression was made in Boston and throughout the country, while at the same time the guilty parties were left to be dealt with in the ordinary course of law, is a striking commentary upon the general peacefulness and decorum of American life, and it shows how high and severe was the standard by which our forefathers judged all lawless proceedings. And here it may not be irrelevant to add that, throughout the constitutional struggles which led to the Revolution, the American standard of political right and wrong was so high that contemporary European politicians found it sometimes difficult to understand it. And for a like reason, even the most fair-minded English historians sometimes fail to see why the Americans should have been so quick to take offence at acts of the British government which doubtless were not meant to be oppressive. If George III. had been a blood-thirsty despot, like Philip II. of Spain; if General Gage had been another Duke of Alva; if American citizens by the hundred had been burned alive or broken on the wheel in New York and Boston; if whole towns had been given up to the cruelty and lust of a beastly soldiery, then no one — not even Dr. Johnson — would have found it hard to understand why the Americans should have exhibited a rebellious temper. But it is one signal characteristic of the progress of political civilization that the part played by sheer brute force in a barbarous age is fully equalled by the part played by a mere covert threat of injustice in a more advanced age. The effect which a blow in the face would produce upon a barbarian will be wrought upon a civilized man by an assertion of some far-reaching legal principle, which only in a subtle and ultimate analysis includes the possibility of a blow in the face. From this point of view, the quickness with which such acts as those of Charles Townshend were comprehended in their remotest bearings is the most striking proof one could wish of the high grade of political culture which our forefathers had reached through their system of perpetual free discussion in town meeting. They had, moreover, reached a point where any manifestation of brute force in the course of a political dispute was exceedingly disgusting and shocking to them. To their minds, the careless slaughter of six citizens conveyed as much meaning as a St. Bartholomew massacre would have conveyed to the minds of men in a lower stage of political development. It was not strange, therefore, that Samuel Adams and his friends should have been ready to make the

Boston Massacre the occasion of a moral lesson to their contemporaries. As far as the poor soldiers were concerned, the most significant fact is that there was no attempt to wreak a paltry vengeance on them. Brought to trial on a charge of murder, after a judicious delay of seven months, they were ably defended by John Adams and Josiah Quincy, and all were acquitted save two, who were convicted of manslaughter, and let off with slight punishment. There were some hotheads who grumbled at the verdict, but the people of Boston generally acquiesced in it, as they showed by immediately choosing John Adams for their representative in the assembly, — a fact which Mr. Lecky calls very remarkable. Such an event as the Boston Massacre could not fail for a long time to point a moral among a people so unused to violence and bloodshed. One of the earliest of American engravers, Paul Revere, published a quaint coloured engraving of the scene in King Street, which for a long time was widely circulated, though it has now become very scarce. At the same time, it was decided that the fatal Fifth of March should be solemnly commemorated each year by an oration to be delivered in the Old South Meeting-House; and this custom was kept up until the recognition of American independence in 1783, when the day for the oration was changed to the Fourth of July.

Lord North's ministry.

Five weeks before the Boston Massacre the Duke of Grafton had resigned, and Lord North had become prime minister of England. The colonies were kept under Hillsborough, and that great friend of arbitrary government, Lord Thurlow, as solicitor-general, became the king's chief legal adviser. George III. was now, to all intents and purposes, his own prime minister, and remained so until after the overthrow at Yorktown. The colonial policy of the government soon became more vexatious than ever. The promised repeal of all the Townshend acts, except the act imposing the tea-duty, was carried through Parliament in April, and its first effect in America, as Lord North had foreseen, was to weaken the spirit of opposition, and to divide the more complaisant colonies from those that were most staunch. The policy of non-importation had pressed with special severity upon the commerce of New York, and the merchants there complained that these fire-eating planters of Virginia and farmers of Massachusetts were growing rich at the expense of their neighbours. In July, the New York merchants broke the non-importation agreement, and sent orders to England for all sorts of merchandise except tea. Such a measure, on the part of so great a seaport, virtually overthrew the non-importation policy, upon which the patriots mainly relied to force the repeal of the Tea Act. The wrath of the other colonies was intense. At the Boston town meeting the letter of the New York merchants was torn in pieces. In New Jersey, the students of Princeton College, James Madison being one of the number, assembled on the green in their black gowns and solemnly burned the letter, while the church-bells were tolled. The offending merchants were stigmatized as

The merchants of New York.

"Revolters," and in Charleston their conduct was vehemently denounced. "You had better send us your old liberty-pole," said Philadelphia to New York, with bitter sarcasm, "for you clearly have no further use for it."

This breaking of the non-importation agreement by New York left no general issue upon which the colonies could be sure to unite unless the ministry should proceed to force an issue upon the Tea Act. For the present, Lord North saw the advantage he had gained, and was not inclined to take any such step. Nevertheless, as just observed, the policy of the government soon became more vexatious than ever. In the summer of 1770, the king entered upon a series of local quarrels with the different colonies, taking care not to raise any general issue. Royal instructions were sent over to the different governments, enjoining courses of action which were unconstitutional and sure to offend the people.

Assemblies convened in strange places.

The assemblies were either dissolved, or convened at strange places, as at Beaufort in South Carolina, more than seventy miles from the capital, or at Cambridge in Massachusetts. The local governments were as far as possible ignored, and local officers were appointed, with salaries to be paid by the Crown. In Massachusetts, these officers were illegally exempted from the payment of taxes.

Taxes in Maryland.

In Maryland, where the charter had expressly provided that no taxes could ever be levied by the British Crown, the governor was ordered to levy taxes indirectly by reviving a law regulating officers' fees, which had expired by lapse of time. In North Carolina, excessive fees were extorted, and the sheriffs in many cases collected taxes of which they rendered no account. The upper counties of both the Carolinas were peopled by a hardy set of small farmers and herdsmen, Presbyterians, of Scotch-Irish pedigree, who were known by the name of "Regulators," because, under the exigencies of their rough frontier life, they formed voluntary associations for the regulation of their own police and the condign punishment of horse-thieves and other criminals. In 1771, the North Carolina Regulators, goaded by repeated acts of extortion and of unlawful imprisonment, rose in rebellion. A fierce battle was fought at Alamance, near the headwaters of the Cape Fear river, in which the Regulators were totally defeated by Governor Tryon, leaving two hundred of their number dead and wounded upon the field: and six of their leaders, taken prisoners, were summarily hanged for treason. For this achievement Tryon was pronounced the ablest of the colonial governors, and was soon promoted to the governorship of New York, where he left his name for a time upon the vaguely defined wilderness beyond Schenectady, known in the literature of the Revolutionary War as Tryon County. The barbarous condition of the frontier where these scenes occurred, and the fact that the militia of the lower counties voluntarily assisted the governor in his campaign against the Regulators, deprived these events of much of the influence they might otherwise have had upon the country; so that it is not

The North Carolina "Regulators."

the Cape Fear but the Concord river that ordinarily occurs to us, when we think of the first bloodshed in the Revolutionary War.

Affair of the Gaspee.

In Rhode Island, the eight-gun schooner Gaspee, commanded by Lieutenant Duddington, was commissioned to enforce the revenue acts along the coasts of Narragansett Bay, and she set about the work with reckless and indiscriminating zeal. "Thorough" was Duddington's motto, as it was Lord Strafford's. He not only stopped and searched every vessel that entered the bay, and seized whatever goods he pleased, whether there was any evidence of their being contraband or not, but, besides this, he stole the sheep and hogs of the farmers near the coast, cut down their trees, fired upon market-boats, and behaved in general with unbearable insolence. In March, 1772, the people of Rhode Island complained of these outrages. The matter was referred to Rear-Admiral Montagu, commanding the little fleet in Boston harbour. Montagu declared that the lieutenant was only doing his duty, and threatened the Rhode Island people in case they should presume to interfere. For three months longer the Gaspee kept up her irritating behaviour, until one evening in June, while chasing a swift American ship, she ran aground. The following night she was attacked by a party of men in eight boats, and captured after a short skirmish, in which Duddington was severely wounded. The crew was set on shore, and the schooner was burned to the water's edge. This act of reprisal was not relished by the government, and large rewards were offered for the arrest of the men concerned in it; but although probably everybody knew who they were, it was impossible to obtain any evidence against them. By a royal order in council, the Rhode Island government was commanded to arrest the offenders and deliver them to Rear-Admiral Montagu, to be taken over to England for trial; but Stephen Hopkins, the venerable chief justice of Rhode Island, flatly refused to take cognizance of any such arrest if made within the colony.

The salaries of the judges.

The black thunder-clouds of war now gathered quickly. In August, 1772, the king ventured upon an act which went further than anything that had yet occurred toward hastening on the crisis. It was ordered that all the Massachusetts judges, holding their places during the king's pleasure, should henceforth have their salaries paid by the Crown, and not by the colony. This act, which aimed directly at the independence of the judiciary, aroused intense indignation. The people of Massachusetts were furious, and Samuel Adams now took a step which contributed more than anything that had yet been done toward organizing the opposition to the king throughout the whole country. The idea of establishing committees of correspondence was not wholly new. The great preacher Jonathan Mayhew had recommended such a step to James Otis in 1766, and he was led to it through his experience of church matters. Writing in haste, on a Sunday morning, he said, "To a good man all time is holy enough; and none is too holy to do good, or to think upon it. Cultivating a good understanding and hearty friendship between these colonies appears to me so necessary a part of prudence and good policy that no favourable opportunity for that purpose should be omitted. . . . You have heard of the *communion of churches:* . . . while I was thinking of this in my bed, the great use and importance of a *communion of colonies* appeared to me in a strong light, which led me immediately to set down these hints to transmit to you." The plan which Mayhew had in mind was the establishment of a regular system of correspondence whereby the colonies could take combined action in defence of their liberties. In the grand crisis of 1772, Samuel Adams saw how much might be effected through committees of correspondence that could not well be effected through the ordinary governmental machinery of the colonies. At the October town meeting in Boston, a committee was appointed to ask the governor whether the judges' salaries were to be paid in conformity to the royal order; and he was furthermore requested to convoke the assembly, in order that the people might have a chance to express their views on so important a matter. But Hutchinson told the committee to mind its own business: he refused to say what would be done about the salaries, and denied the right of the town to petition for a meeting of the assembly. Massachusetts was thus virtually without a general government at a moment when the public mind was agitated by a question of supreme importance. Samuel Adams thereupon in town meeting moved the appointment of a committee of

Jonathan Mayhew's suggestion.

correspondence, "to consist of twenty-one persons, to state the rights of the colonists and of this province in particular, as men and Christians and as subjects; and to communicate and publish the same to the several towns and to the world as the sense of this town, with the infringements and violations thereof that have been, or from time to time may be, made." The adoption of this measure at first excited the scorn of Hutchinson, who described the committee as composed of "deacons," "atheists," and "black-hearted fellows," whom one would not care to meet in the dark. He predicted that they would only make themselves ridiculous, but he soon found reason to change his mind. The response to the statements of the Boston committee was prompt and unanimous, and before the end of the year more than eighty towns had already organized their committees of correspondence. (Here was a new legislative body, springing directly from the people, and competent, as events soon showed, to manage great affairs. Its influence reached into every remotest corner of Massachusetts, it was always virtually in session, and no governor could dissolve or prorogue it. Though unknown to the law, the creation of it involved no violation of law. The right of the towns of Massachusetts to ask one another's advice could no more be disputed than the right of the freemen of any single town to hold a town meeting. The power thus created was omnipresent, but intangible.) "This," said Daniel Leonard, the great Tory pamphleteer, two years afterwards, "is the foulest, subtlest, and

The committees of correspondence in Massachusetts.

most venomous serpent ever issued from the egg of sedition. It is the source of the rebellion. I saw the small seed when it was planted: it was a grain of mustard. I have watched the plant until it has become a great tree. The vilest reptiles that crawl upon the earth are concealed at the root; the foulest birds of the air rest upon its branches. I would now induce you to go to work immediately with axes and hatchets and cut it down, for a twofold reason, — because it is a pest to society, and lest it be felled suddenly by a stronger arm, and crush its thousands in its fall."

Intercolonial committees of correspondence.

The system of committees of correspondence did indeed grow into a mighty tree; for it was nothing less than the beginning of the American Union. Adams himself by no means intended to confine his plan to Massachusetts, for in the following April he wrote to Richard Henry Lee of Virginia urging the establishment of similar committees in every colony. But Virginia had already acted in the matter. When its assembly met in March, 1773, the news of the refusal of Hopkins to obey the royal order, of the attack upon the Massachusetts judiciary, and of the organization of the committees of correspondence was the all-exciting subjects of conversation. The motion to establish a system of intercolonial committees of correspondence was made by the youthful Dabney Carr, and eloquently supported by Patrick Henry and Richard Henry Lee. It was unanimously adopted, and very soon several other colonies elected committees, in response to the invitation from Virginia.

This was the most decided step toward revolution that had yet been taken by the Americans. It only remained for the various intercolonial committees to assemble together, and there would be a Congress speaking in the name of the continent. To bring about such an act of union, nothing more was needed than some fresh course of aggression on the part of the British government which should raise a general issue in all the colonies; and, with the rare genius for blundering which had possessed it ever since the accession of George III., the government now went on to provide such an issue. It was preëminently a moment when the question of taxation should have been let alone. Throughout the American world there was a strong feeling of irritation, which might still have been allayed had the ministry shown a yielding temper. The grounds of complaint had come to be different in the different colonies, and in some cases, in which we can clearly see the good sense of Lord North prevailing over the obstinacy of the king, the ministry had gained a point by yielding. In the Rhode Island case, they had seized a convenient opportunity and let the matter drop, to the manifest advantage of their position. In Massachusetts, the discontent had come to be alarming, and it was skilfully organized. The assembly had offered the judges their salaries in the usual form, and had threatened to impeach them if they should dare to accept a penny from the Crown. The recent action of Virginia had shown that these two most powerful of the colonies were in strong sympathy

The question of taxation revived.

3795

with one another. It was just this moment that George III. chose for reviving the question of taxation, upon which all the colonies would be sure to act as a unit, and sure to withstand him to his face. The duty on tea had been retained simply as a matter of principle. It did not bring three hundred pounds a year into the British exchequer. But the king thought this a favourable time for asserting the obnoxious principle which the tax involved.

Thus, as in Mrs. Gamp's case, a teapot became the cause or occasion of a division between friends. The measures now taken by the government brought matters at once to a crisis. None of the colonies would take tea on its terms. Lord Hillsborough had lately been superseded as colonial secretary by Lord Dartmouth, an amiable man like the prime minister, but like him wholly under the influence of the king. Lord Dartmouth's appointment was made the occasion of introducing a series of new measures. The affairs of the East India Company were in a bad condition, and it was thought that the trouble was partly due to the loss of the American trade in tea. The Americans would not buy tea shipped from England, but they smuggled it freely from Holland, and the smuggling could not be stopped by mere force. The best way to obviate the difficulty, it was thought, would be to make English tea cheaper in America than foreign tea, while still retaining the duty of threepence on a pound. If this could be achieved, it was supposed that the Americans would be sure to buy English tea by reason of its cheapness, and would thus be ensnared into admitting the principle involved in the duty. This ingenious scheme shows how unable the king and his ministers were to imagine that the Americans could take a higher view of the matter than that of pounds, shillings, and pence. In order to enable the East India Company to sell its tea cheap in America, a drawback was allowed of all the duties which such tea had been wont to pay on entering England on its way from China. In this way, the Americans would now find it actually cheaper to buy the English tea with the duty on it than to smuggle their tea from Holland. To this scheme, Lord North said, it was of no use for any one to offer objections, for the king would have it so. "The king meant to try the question with America." In accordance with this policy, several ships loaded with tea set sail in the autumn of 1773 for the four principal ports, Boston, New York, Philadelphia, and Charleston. Agents or consignees of the East India Company were appointed by letter to receive the tea in these four towns.

The king's ingenious scheme.

As soon as the details of this scheme were known in America, the popular wrath was even greater than that which had been stirred up by the Stamp Act, and the whole country was at once in a blaze, from Maine to Georgia. Nevertheless, only legal measures of resistance were contemplated. In Philadelphia, a great meeting was held in October at the State House, and it was voted that whosoever should lend countenance to the receiving or unloading of the tea would be regarded as an enemy to his country. The consignees were then requested to resign their commissions, and did so. In New York and Charleston, also, the consignees threw up their commissions. In Boston, a similar demand was made, but the consignees doggedly refused to resign; and thus the eyes of the whole country were directed toward Boston as the battlefield on which the great issue was to be tried.

How Boston became the battle-ground.

During the month of November many town meetings were held in Faneuil Hall. On the 17th, authentic intelligence was brought that the tea-ships would soon arrive. The next day, a committee, headed by Samuel Adams, waited upon the consignees, and again asked them to resign. Upon their refusal, the town meeting instantly dissolved itself, without a word of comment or debate; and at this ominous silence the consignees and the governor were filled with a vague sense of alarm, as if some storm were brewing whereof none could foresee the results. All felt that the decision now rested with the committees of correspondence. Four days afterward, the committees of Cambridge, Brookline, Roxbury, and Dorchester met the Boston committee at Faneuil Hall, and it was unanimously resolved that on no account should the tea be landed. The five towns also sent a letter to all the other towns in the colony, saying, "Brethren, we are reduced to this dilemma: either to sit down quiet under this and every other burden that our enemies shall see fit to lay upon us, or to rise up and resist this and every plan laid for our destruction, as becomes

The five towns ask advice.

wise freemen. In this extremity we earnestly request your advice." There was nothing weak or doubtful in the response. From Petersham and Lenox perched on their lofty hilltops, from the valleys of the Connecticut and the Merrimack, from Chatham on the bleak peninsula of Cape Cod, there came but one message, — to give up life and all that makes life dear, rather than submit like slaves to this great wrong. Similar words of encouragement came from other colonies. In Philadelphia, at the news of the bold stand Massachusetts was about to take, the church-bells were rung, and there was general rejoicing about the streets. A letter from the men of Philadelphia to the men of Boston said, "Our only fear is lest you may shrink. May God give you virtue enough to save the liberties of your country."

Arrival of the tea; meeting at the Old South.

On Sunday, the 28th, the Dartmouth, first of the tea-ships, arrived in the harbour. The urgency of the business in hand overcame the sabbatarian scruples of the people. The committee of correspondence met at once, and obtained from Rotch, the owner of the vessel, a promise that the ship should not be entered before Tuesday. Samuel Adams then invited the committees of the five towns, to which Charlestown was now added, to hold a mass-meeting the next morning at Faneuil Hall. More than five thousand people assembled, but as the Cradle of Liberty could not hold so many, the meeting was adjourned to the Old South Meeting-House. It was voted, without a single dissenting voice, that the tea should be sent back to England in the ship which had brought it. Rotch was forbidden to enter the ship at the Custom House, and Captain Hall, the ship's master, was notified that "it was at his peril if he suffered any of the tea brought by him to be landed." A night-watch of twenty-five citizens was set to guard the vessel, and so the meeting adjourned till next day, when it was understood that the consignees would be ready to make some proposals in the matter. Next day, the message was brought from the consignees that it was out of their power to send back the tea; but if it should be landed, they declared themselves willing to store it, and not expose any of it for sale until word could be had from England. Before action could be taken upon this message, the sheriff of Suffolk county entered the church and read a proclamation from the governor, warning the people to disperse and "surcease all further unlawful proceedings at their utmost peril." A storm of hisses was the only reply, and the business of the meeting went on. The proposal of the consignees was rejected, and Rotch and Hall, being present, were made to promise that the tea should go back to England in the Dartmouth, without being landed or paying duty. Resolutions were then passed, forbidding all owners or masters of ships to bring any tea from Great Britain to any part of Massachusetts, so long as the act imposing a duty on it remained unrepealed. Whoever should disregard this injunction would be treated as an enemy to his country, his ships would be prevented from landing — by force, if necessary — and his tea would be sent back to the place whence it came. It was further voted that the citizens of Boston and the other towns here assembled would see that these resolutions were carried into effect, "at the risk of their lives and property." Notice of these resolutions was sent to the owners of the other ships, now daily expected. And, to crown all, a committee, of which Adams was chairman, was appointed to send a printed copy of these proceedings to New York and Philadelphia, to every seaport in Massachusetts, and to the British government.

Two or three days after this meeting, the other two ships arrived, and, under orders from the committee of correspondence, were anchored by the side of the Dartmouth, at Griffin's Wharf, near the foot of Pearl Street. A military watch was kept at the wharf day and night, sentinels were placed in the church belfries, chosen post-riders, with horses saddled and bridled, were ready to alarm the neighbouring towns, beacon-fires were piled all ready for lighting upon every hilltop, and any attempt to land the tea forcibly would have been the signal for an instant uprising throughout at least four counties. Now, in accordance with the laws providing for the entry and clearance of shipping at custom houses, it was necessary that every ship should land its cargo within twenty days from its arrival. In case this was not done, the revenue officers were authorized to seize the ship and land its cargo themselves. In the case of the Dartmouth, the captain had promised to take her back to England without unloading; but still, before she could

The tea-ships placed under guard.

legally start, she must obtain a clearance from the collector of customs, or, in default of this, a pass from the governor. At sunrise of Friday, the 17th of December, the twenty days would have expired.

On Saturday, the 11th, Rotch was summoned before the committee of correspondence, and Samuel Adams asked him why he had not kept his promise, and started his ship off for England. He sought to excuse himself on the ground that he had not the power to do so, whereupon he was told that he must apply to the collector for a clearance. Hearing of these things, the governor gave strict orders at the Castle to fire upon any vessel trying to get out to sea without a proper permit; and two ships from Montagu's fleet, which had been laid up for the winter, were stationed at the entrance of the harbour, to make sure against the Dartmouth's going out. Tuesday came, and Rotch, having done nothing, was summoned before the town meeting, and peremptorily ordered to apply for a clearance. Samuel Adams and nine other gentlemen accompanied him to the Custom House to witness the proceedings, but the collector refused to give an answer until the next day. The meeting then adjourned till Thursday, the last of the twenty days. On Wednesday morning, Rotch was again escorted to the Custom House, and the collector refused to give a clearance unless the tea should first be landed.

On the morning of Thursday, December 16th, the assembly which was gathered in the Old South Meeting-House, and in the streets about it, numbered more than seven thousand people. It was to be one of the most momentous days in the history of the world. The clearance having been refused, nothing now remained but to order Rotch to request a pass for his ship from the governor. But the wary Hutchinson, well knowing what was about to be required of him, had gone out to his country house at Milton, so as to foil the proceedings by his absence. But the meeting was not to be so trifled with. Rotch was enjoined, on his peril, to repair to the governor at Milton, and ask for his pass; and while he was gone, the meeting considered what was to be done in case of a refusal. Without a pass it would be impossible for the ship to clear the harbour under the guns of the Castle; and by sunrise, next morning, the revenue officers would be empowered to seize the ship, and save by a violent assault upon them it would be impossible to prevent the landing of the tea. "Who knows," said John Rowe, "how tea will mingle with salt water?" And great applause followed the suggestion. Yet the plan which was to serve as a last resort had unquestionably been adopted in secret committee long before this. It appears to have been worked out in detail in a little back room at the office of the "Boston Gazette," and there is no doubt that Samuel Adams, with some others of the popular leaders, had a share in devising it. But among the thousands present at the town meeting, it is probable that very few knew just what it was designed to do. At five in the afternoon, it was unanimously voted that, come what would, the tea should not be landed. It had now grown dark,

Town meeting at the Old South.

and the church was dimly lighted with candles. Determined not to act until the last legal method of relief should have been tried and found wanting, the great assembly was still waiting quietly in and about the church when, an hour after nightfall, Rotch returned from Milton with the governor's refusal. Then, amid profound stillness, Samuel Adams arose and said, quietly but distinctly, "This meeting can do nothing more to save the country." It was the declaration of war; the law had shown itself unequal to the occasion, and nothing now remained but a direct appeal to force. Scarcely had the watchword left his mouth when a war-whoop answered from outside the door, and fifty men in the guise of Mohawk Indians passed quickly by the entrance, and hastened to Griffin's Wharf. Before the nine o'clock bell rang, the three hundred and forty-two chests of tea laden upon the three ships had been cut open, and their contents emptied into the sea. Not a person was harmed; no other property was injured; and the vast crowd, looking upon the scene from the wharf in the clear frosty moonlight, was so still that the click of the hatchets could be distinctly heard. Next morning, the salted tea, as driven by wind and wave, lay in long rows on Dorchester beach, while Paul Revere, booted and spurred, was riding post-haste to Philadelphia, with the glorious news that Boston had at last thrown down the gauntlet for the king of England to pick up.

The tea thrown into the harbour.

This heroic action of Boston was greeted with public rejoicing throughout all the thirteen colonies, and the other principal seaports were not slow to follow the example. A ship laden with two hundred and fifty-seven chests of tea had arrived at Charleston on the 2d of December; but the consignees had resigned, and after twenty days the ship's cargo was seized and landed; and so, as there was no one to receive it, or pay the duty, it was thrown into a damp cellar, where it spoiled. In Philadelphia, on the 25th, a ship arrived with tea; but a meeting of five thousand men forced the consignees to resign, and the captain straightway set sail for England, the ship having been stopped before it had come within the jurisdiction of the custom house.

In Massachusetts, the exultation knew no bounds. "This," said John Adams, "is the most magnificent movement of all. There is a dignity, a majesty, a sublimity, in this last effort of the patriots that I greatly admire." Indeed, often as it has been cited and described, the Boston Tea Party was an event so great that even American historians have generally failed to do it justice. This supreme assertion by a New England town meeting of the most fundamental principle of political freedom has been curiously misunderstood by British writers, of whatever party. The most recent Tory historian, Mr. Lecky,[1] speaks of "the Tea-riot at Boston," and characterizes it as an "outrage." The most recent Liberal historian, Mr. Green, alludes to it as "a trivial riot." Such expressions betray most

Grandeur of the Boston Tea Party.

[1] In his account of the American Revolution Mr. Lecky inclines to the Tory side, but he is eminently fair and candid.

profound misapprehension alike of the significance of this noble scene and of the political conditions in which it originated. There is no difficulty in defining a riot. The pages of history teem with accounts of popular tumults, wherein passion breaks loose and wreaks its fell purpose, unguided and unrestrained by reason. No definition could be further from describing the colossal event which occurred in Boston on the 16th of December, 1773. Here passion was guided and curbed by sound reason at every step, down to the last moment, in the dim candle-light of the old church, when the noble Puritan statesman quietly told his hearers that the moment for using force had at last, and through no fault of theirs, arrived. They had reached a point where the written law had failed them; and in their effort to defend the eternal principles of natural justice, they were now most reluctantly compelled to fall back upon the paramount law of self-preservation. It was the one supreme moment in a controversy supremely important to mankind, and in which the common sense of the world has since acknowledged that they were wholly in the right. It was the one moment of all that troubled time in which no compromise was possible. "Had the tea been landed," says the contemporary historian Gordon, "the union of the colonies in opposing the ministerial scheme would have been dissolved; and it would have been extremely difficult ever after to have restored it." In view of the stupendous issues at stake, the patience of the men of Boston was far more remarkable than their boldness. For the quiet sublimity of reasonable but dauntless moral purpose, the heroic annals of Greece and Rome can show us no greater scene than that which the Old South Meeting-House witnessed on the day when the tea was destroyed.

When the news of this affair reached England, it was quite naturally pronounced by Lord North a fitting culmination to years of riot and lawlessness. This, said Lord George Germain, is what comes of their wretched old town meetings. The Americans have really no government. These "are the proceedings of a tumultuous and riotous rabble, who ought, if they had the least prudence, to follow their mercantile employments, and not trouble themselves with politics and government, which they do not understand. Some gentlemen say, 'Oh, don't break their charter; don't take away rights granted them by the predecessors of the Crown.' Whoever wishes to preserve such charters, I wish him no worse than to govern such subjects." "These remarks," said Lord North, "are worthy of a great mind." "If we take a determined stand now," said Lord Mansfield, "Boston will submit, and all will end in victory without carnage." "The town of Boston," said Mr. Venn, "ought to be knocked about their ears and destroyed. You will never meet with proper obedience to the laws of this country until you have destroyed that nest of locusts." General Gage, who had just come home on a visit, assured the king that the other colonies might speak fair words to Massachusetts, but would do nothing to help her; and he offered with four regiments to make a speedy end of the whole

How Parliament received the news.

matter. "They will be lions," said Gage, "while we are lambs; but if we take the resolute part, they will prove very meek, I promise you." It was in this spirit and under the influence of these ideas that the ministry took up the business of dealing with the refractory colony of Massachusetts. Lord North proposed a series of five measures, which, from the king's point of view, would serve, not only to heal the wounded pride of Great Britain, but also to prevent any more riotous outbreaks among this lawless American people. Just at this moment, the opposition ventured upon a bold stroke. Fox said truly that no plan for pacifying the colonies would be worth a rush unless the unconditional repeal of the Tea Act should form part of it. A bill for the repealing of the Tea Act was brought in by Fuller, and a lively debate ensued, in the course of which Edmund Burke made one of the weightiest speeches ever heard in the House of Commons; setting forth in all the wealth of his knowledge the extreme danger of the course upon which the ministry had entered, and showing how little good fruit was to be expected from a coercive policy, even if successful. Burke was ably supported by Fox, Conway, Barré, Savile, Dowdeswell, Pownall, and Dunning. But the current had set too strongly against conciliation. Lord North sounded the keynote of the whole British policy when he said, "To repeal the tea-duty would stamp us with timidity." Come what might, it would never do for the Americans to get it into their heads that the government was not all-powerful. They must be humbled first, that they might be reasoned with afterwards. The tea-duty, accordingly, was not repealed, but Lord North's five acts for the better regulation of American affairs were all passed by Parliament.

By the first act, known as the Boston Port Bill, no ships were to be allowed to enter or clear the port of Boston until the rebellious town should have indemnified the East India Company for the loss of its tea, and should otherwise have made it appear to the king that it would hereafter show a spirit of submission. Marblehead was made a port of entry instead of Boston, and Salem was made the seat of government.

The Boston Port Bill.

By the second act, known as the Regulating Act, the charter of Massachusetts was annulled without preliminary notice, and her free government was destroyed. Under the charter, the members of the council for each year were chosen in a convention consisting of the council of the preceding year and the assembly. Each councillor held office for a year, and was paid out of an appropriation made by the assembly. Now, hereafter, the members of the council were to be appointed by the governor on a royal writ of *mandamus*, their salaries were to be paid by the Crown, and they could be removed from office at the king's pleasure. The governor was empowered to appoint all judges and officers of courts, and all such officers were to be paid by the king and to hold office during his pleasure. The governor and his dependent council could appoint sheriffs and remove them without assigning any reason, and these dependent sheriffs were to have the sole right of

The Regulating Act.

returning juries. But, worse than all, the town-meeting system of local self-government was ruthlessly swept away. Town meetings could indeed be held twice a year for the election of town officers, but no other business could be transacted in them. The effect of all these changes would, of course, be to concentrate all power in the hands of the governor, leaving no check whatever upon his arbitrary will. It would, in short, transform the free commonwealth of Massachusetts into an absolute despotism, such as no Englishman had ever lived under in any age. And this tremendous act was to go into operation on the first day of the following June.

By the third act — a pet measure of George III., to which Lord North assented with great reluctance — it was provided that if any magistrate, soldier, or revenue officer in Massachusetts should be indicted for murder, he should be tried, not in Massachusetts, but in Great Britain. This measure — though doubtless unintentionally — served to encourage the soldiery in shooting down peaceful citizens, and it led by a natural sequence to the bloodshed on Lexington green. It was defended on the ground that in case of any chance affray between soldiers and citizens, it would not be possible for the soldiers to obtain a fair trial in Massachusetts. Less than four years had elapsed since Preston's men had been so readily acquitted of murder after the shooting in King Street, but such facts were of no avail now. The momentous bill passed in the House of Commons by a vote of more than four to one, in spite of Colonel Barré's ominous warnings.

The shooting of citizens.

By the fourth act all legal obstacles to the quartering of troops in Boston or any other town in Massachusetts were swept away.

By the fifth act, known as the Quebec Act, the free exercise of the Catholic religion was sanctioned throughout Canada, — a very judicious measure of religious toleration, which concerned the other colonies but little, however it might in some cases offend their prejudices. But this act went on to extend the boundaries of Canada southward to the Ohio river, in defiance of the territorial claims of Massachusetts, Connecticut, New York, and Virginia. This extensive region, the part of North America which was next to be colonized by men of English race, was to be governed by a viceroy, with despotic powers; and such people as should come to live there were to have neither popular meetings, nor *habeas corpus*, nor freedom of the press. "This," said Lord Thurlow, "is the only sort of constitution fit for a colony," — and all the American colonies, he significantly added, had better be reduced to this condition as soon as possible.

The Quebec Act.

When all these acts had been passed, in April, 1774, General Gage was commissioned to supersede Hutchinson temporarily as governor of Massachusetts, and was sent over with as little delay as possible, together with the four regiments which were to scare the people into submission. On the first day of June, he was to close the port of Boston and begin starving the town into good behaviour; he was to arrest the leading patriots and send them to England for trial; and

Gage sent to Boston.

he was expressly authorized to use his own discretion as to allowing the soldiers to fire upon the people. All these measures for enslaving peaceful and law-abiding Englishmen the king of England now contemplated, as he himself declared, "with supreme satisfaction."

In recounting such measures as these, the historian is tempted to pause for a moment, and ask whether it could really have been an *English* government that planned and decreed such things. From the autocratic mouth of an Artaxerxes or an Abderrahman one would naturally expect such edicts to issue. From the misguided cabinets of Spain and France, in evil times, measures in spirit like these had been known to proceed. But England had for ages stood before the world as the staunch defender of personal liberty and of local self-government; and through the mighty strength which this spirit of freedom, and nothing else, had given her, she had won the high privilege of spreading her noble and beneficent political ideas over the best part of the habitable globe. Yet in the five acts of this political tragedy of 1774 we find England arrayed in hostility to every principle of public justice which Englishmen had from time immemorial held sacred. Upon the great continent which she had so lately won from the French champions of despotism, we see her, in a fit of obstinate anger, vainly seeking to establish a tyrannical *régime* no better than that which but yesterday it had been her glory to overthrow. Such was the strange, the humiliating, the self-contradictory attitude into which England had at length been brought by the short-sighted Tory policy of George III.!

But this policy was no less futile than it was unworthy of the noble, freedom-loving English people. For after that fated 1st of June, the sovereign authority of Great Britain, whether exerted through king or through Parliament, was never more to be recognized by the men of Massachusetts.

CHAPTER III.

THE CONTINENTAL CONGRESS.

THE unfortunate measures of April, 1774, were not carried through Parliament without earnest opposition. Lord Rockingham and his friends entered a protest on the journal of the House of Lords, on the grounds that the people of Massachusetts had not been heard in their own defence, and that the lives and liberties of the citizens were put absolutely into the hands of the governor and council, who were thus invested with greater powers than it had ever been thought wise to entrust to the king and his privy council in Great Britain. They concluded, therefore, that the acts were unconstitutional. The Duke of Richmond could not restrain his burning indignation. "I wish," said he in the House of Lords, — "I wish from the bottom of my heart that the Americans may resist, and get the better of the forces sent against them." But that the Americans really would resist, very few people in England believed. The conduct of the ministry was based throughout upon the absurd idea that the Americans could be frightened into submission. General Gage, as we have seen, thought that four regiments would be enough to settle the whole business. Lord Sandwich said that

Protests of the Whigs.

the Americans were a set of undisciplined cowards, who would take to their heels at the first sound of a cannon. Even Hutchinson, who went over to England about this time, and who ought to have known of what stuff the men of Massachusetts were made, assured the king that they never would resist a regular army. Such blunders, however, need not surprise us when we recollect how, just before the war of secession, the people of the southern and of the northern states made similar mistakes with regard to each other. In 1860, it was commonly said by Southern people that Northern people would submit to anything rather than fight; and in support of this opinion, it was sometimes asked, "If the Northern people are not arrant cowards, why do they never have duels?" On the other hand, it was commonly said at the North that the Southern people, however bravely they might bluster, would never enter upon a war of secession, because it was really much more for their interest to remain in the Federal Union than to secede from it, — an argument which lost sight of one of the commonest facts in human life, that under the influence of strong passion men are unable to take just views of what concerns their own interests. Such examples show how hard it often is for one group of men to understand another group, even when they are all of the same blood and speech, and think alike about most matters that do not touch the particular subject in dispute. Nothing could have been surer, either in 1860 or 1774, than that the one party to the quarrel was as bold and brave as the other.

Belief that the Americans would not fight.

Another fatal error under which the ministry laboured was the belief that Massachusetts would not be supported by the other colonies. Their mistake was not unlike that which ruined the plans of Napoleon III., when he declared war upon Prussia in 1870.

Belief that Massachusetts would not be supported by the other colonies.

There was no denying the fact of strong jealousies among the American colonies in 1774, as there was no denying the fact of strong jealousies between the northern and southern German states in 1870. But the circumstances under which Napoleon III. made war on Prussia happened to be such as to enlist all the German states in the common cause with her. And so it was with the war of George III. against Massachusetts. As soon as the charter of that colony was annulled, all the other colonies felt that their liberties were in jeopardy; and thence, as Fox truly said, "all were taught to consider the town of Boston as suffering in the common cause."

News of the Boston Port Bill was received in America on the 10th of May. On the 12th the committees of several Massachusetts towns held a convention at Faneuil Hall, and adopted a circular letter, prepared by Samuel Adams, to be sent to all the other colonies, asking for their sympathy and coöperation. The response was prompt and emphatic. In the course of the summer, conventions were held in nearly all the colonies, declaring that Boston should be regarded as "suffering in the common cause." The obnoxious acts of Parliament were printed on paper with deep black borders, and in some towns

News of the Port Bill.

were publicly burned by the common hangman. Droves of cattle and flocks of sheep, cartloads of wheat and maize, kitchen vegetables and fruit, barrels of sugar, quintals of dried fish, provisions of every sort, were sent overland as free gifts to the people of the devoted city, even the distant rice-swamps of South Carolina contributing their share. The over-cautious Franklin had written from London, suggesting that perhaps it might be best, after all, for Massachusetts to indemnify the East India Company; but Gadsden, with a sounder sense of the political position, sent word, "Don't pay for an ounce of the damned tea." Throughout the greater part of the country the 1st of June was kept as a day of fasting and prayer; bells were muffled and tolled in the principal churches; ships in the harbours put their flags at half-mast. Marblehead, which was appointed to supersede Boston as port of entry, immediately invited the merchants of Boston to use its wharfs and warehouses free of charge in shipping and unshipping their goods. A policy of absolute non-importation was advocated by many of the colonies, though Pennsylvania, under the influence of Dickinson, still vainly cherishing hopes of reconciliation, hung back, and advised that the tea should be paid for. As usual, the warmest sympathy with New England came from Virginia. "If need be," said Washington, "I will raise one thousand men, subsist them at my own expense, and march myself at their head for the relief of Boston."

To insure concerted action on the part of the

whole country, something more was required than these general expressions and acts of sympathy. The proposal for a Continental Congress came first from the Sons of Liberty in New York; it was immediately taken up by the members of the Virginia legislature, sitting in convention at the Raleigh tavern, after the governor had dissolved them as a legislature, and Massachusetts was invited to appoint the time and place for the meeting of the Congress. On the 7th of June the Massachusetts assembly was convened at Salem by General Gage, in conformity with the provisions of the Port Bill. Samuel Adams always preferred to use the ordinary means of transacting public business so long as they were of avail, and he naturally wished to have the act appointing a Continental Congress passed by the assembly. But this was not easy to bring about, for upon the first hint that any such business was to come up the governor would instantly dissolve the assembly. In such case it would be necessary for the committees of correspondence throughout Massachusetts to hold a convention for the purpose of appointing the time and place for the Congress and of electing delegates to attend it. But Adams preferred to have these matters decided in regular legislative session, and he carried his point. Having talked privately with several of the members, at last on the 17th of June — a day which a twelvemonth hence was to become so famous — the favourable moment came. Having had the door locked, he introduced his resolves, appointing five delegates to confer with

Samuel Adams at Salem.

duly appointed delegates from the other colonies, in a Continental Congress at Philadelphia on the 1st of September next. Some of the members, astonished and frightened, sought to pass out; and as the doorkeeper seemed uneasy at assuming so much responsibility, Samuel Adams relieved him of it by taking the key from the door and putting it into his own pocket, whereupon the business of the assembly went on. Soon one of the Tory members pretended to be very sick, and being allowed to go out, made all haste to Governor Gage, who instantly drew up his writ dissolving the assembly, and sent his secretary with it. When the secretary got there, he found the door locked, and as nobody would let him in or pay any attention to him, he was obliged to content himself with reading the writ, in a loud voice, to the crowd which had assembled on the stairs. The assembly meanwhile passed the resolves by 117 to 12, elected Samuel and John Adams, Thomas Cushing, and Robert Treat Paine as delegates, assessed the towns in the commonwealth for the necessary expenses, passed measures for the relief of Boston, and adjourned *sine die*. All the other colonies except Georgia, in the course of the summer, accepted the invitation, and chose delegates, either through their assemblies or through special conventions. Georgia sent no delegates, but promised to adopt any course of action that should be determined upon.

Before the time appointed for the Congress, Massachusetts had set the Regulating Act at defiance. On the 16th of August, when the court assembled at Great Barrington, a vast multitude of farmers surrounded the court-house and forbade the judges to transact any business. Two or three of the councillors newly appointed on the king's writ of *mandamus* yielded in advance to public opinion, and refused to take their places. Those who accepted were forced to resign. At Worcester 2,000 men assembled on the common, and compelled Timothy Paine to make his resignation in writing. The councillor appointed from Bridgewater was a deacon; when he read the psalm the congregation refused to sing. In Plymouth one of the most honoured citizens, George Watson, accepted a place on the council; as he took his seat in church on the following Sunday, the people got up and began to walk out of the house. Overcome with shame, for a moment his venerable gray head sank upon the pew before him; then he rose up and vowed that he would resign. In Boston the justices and barristers took their accustomed places in the court-house, but no one could be found to serve as juror in a court that was illegally constituted. Gage issued a proclamation warning all persons against attending town-meeting, but no one heeded him, and town-meetings were more fully attended than ever. He threatened to send an armed force against Worcester, but the people there replied that he would do so at his peril, and forthwith began to collect powder and ball. At Salem the people walked to the town-house under the governor's nose and in the very presence of a line of soldiers. On the 1st of September a party

Massachusetts nullifies the Regulating Act.

of soldiers seized two hundred kegs of powder at Charlestown and two field-pieces at Cambridge, and carried them to Castle William. As the news spread about the country, rumour added that the troops had fired upon the people, and within forty-eight hours at least 20,000 men were marching on Boston; but they turned back to their homes on receiving word from the Boston committee that their aid was not yet needed.

John Hancock and Joseph Warren.

During these stirring events, in the absence of Samuel Adams, who had gone to attend the Congress at Philadelphia, the most active part in the direction of affairs at Boston was taken by Dr. Joseph Warren. This gentleman — one of a family which has produced three very eminent physicians — was graduated at Harvard College in 1759. He had early attracted the attention of Samuel Adams, had come to be one of his dearest friends, and had been concerned with him in nearly all of his public acts of the past seven years. He was a man of knightly bravery and courtesy, and his energy and fertility of mind were equalled only by his rare sweetness and modesty. With Adams and Hancock, he was one of the great Massachusetts triumvirate of Revolutionary leaders. The accession of Hancock to the Revolutionary cause at an early period had been of great help, by reason of his wealth and social influence. Hancock was graduated at Harvard College in 1754. He was a gentleman of refinement and grace, but neither for grasp of intelligence nor for strength of character can he be compared with Adams or with Warren.

His chief weakness was personal vanity, but he was generous and loyal, and under the influence of the iron-willed Adams was capable of good things. Upon Warren, more than any one else, however, Adams relied as lieutenant, who, under any circumstances whatever, would be sure to prove equal to the occasion.

On the 5th of September Gage began fortifying Boston Neck, so as to close the only approach to the city by land. Next day the county assize was to be held at Worcester; but 5,000 armed men, drawn up in regular military array, lined each side of the main street, and the unconstitutionally appointed judges were forbidden to take their seats. On the same day a convention of the towns of Suffolk County was held at Milton, and a series of resolutions, drawn up by Dr. Warren, were adopted unanimously. The resolutions declared that a king who violates the chartered rights of his people forfeits their allegiance; they declared the Regulating Act null and void, and ordered all the officers appointed under it to resign their offices at once; they directed the collectors of taxes to refuse to pay over money to Gage's treasurer; they advised the towns to choose their own militia officers; and they threatened the governor that, should he venture to arrest any one for political reasons, they would retaliate by seizing upon the Crown officers as hostages. A copy of these resolutions, which virtually placed Massachusetts in an attitude of rebellion, was forwarded to the Continental Congress, which enthusiastically indorsed them, and pledged

The Suffolk County Resolves, Sept. 6, 1774.

the faith of all the other colonies that they would aid Massachusetts in case armed resistance should become inevitable, while at the same time they urged that a policy of moderation should be preserved, and that Great Britain should be left to fire the first shot.

On receiving these instructions from the Congress, the people of Massachusetts at once proceeded to organize a provisional government in accordance with the spirit of the Suffolk resolves. Gage had issued a writ convening the assembly at Salem for the 1st of October, but before the day arrived he changed his mind, and prorogued it. In disregard of this order, however, the representatives met at Salem a week later, organized themselves into a provincial congress, with John Hancock for president, and adjourned to Concord. On the 27th they chose a committee of safety, with Warren for chairman, and charged it with the duty of collecting military stores. In December this Congress dissolved itself, but a new one assembled at Cambridge on the 1st of February, and proceeded to organize the militia and appoint general officers. A special portion of the militia, known as "minute-men," were set apart, under orders to be ready to assemble at a moment's warning; and the committee of safety were directed to call out this guard as soon as Gage should venture to enforce the Regulating Act. Under these instructions every village green in Massachusetts at once became the scene of active drill. Nor was it a population unused to arms that thus began to marshal itself into

Provincial Congress in Massachusetts.

companies and regiments. During the French war one fifth of all the able-bodied men of Massachusetts had been in the field, and in 1757 the proportion had risen to one third. There were plenty of men who had learned how to stand under fire, and officers who had held command on hard-fought fields; and all were practised marksmen. It is quite incorrect to suppose that the men who first repulsed the British regulars in 1775 were a band of farmers, utterly unused to fighting. Their little army was indeed a militia, but it was made up of warlike material.

Meeting of the Continental Congress, Sept. 5, 1774.

While these preparations were going on in Massachusetts, the Continental Congress had assembled at Philadelphia on the 5th of September. Peyton Randolph, of Virginia, was chosen president; and the Adamses, the Livingstons, the Rutledges, Dickinson, Chase, Pendleton, Lee, Henry, and Washington took part in the debates. One of their first acts was to dispatch Paul Revere to Boston with their formal approval of the action of the Suffolk Convention. After four weeks of careful deliberation they agreed upon a declaration of rights, claiming for the American people "a free and exclusive power of legislation in their provincial legislatures, where their rights of legislation could alone be preserved in all cases of taxation and internal polity." This paper also specified the rights of which they would not suffer themselves to be deprived, and called for the repeal of eleven acts of Parliament by which these rights had been infringed. Besides this, they formed an association for insuring commercial non-intercourse with Great Britain, and charged the committees of correspondence with the duty of inspecting the entries at all custom-houses. Addresses were also prepared, to be sent to the king, to the people of Great Britain, and to the inhabitants of British America. The 10th of May was appointed for a second Congress, in which the Canadian colonies and the Floridas were invited to join; and on the 26th of October the Congress dissolved itself.

The ability of the papers prepared by the first Continental Congress has long been fully admitted in England as well as in America. Chatham declared them unsurpassed by any state papers ever composed in any age or country. But Parliament was not now in the mood for listening to reason. Chatham, Shelburne, and Camden urged in vain that the vindictive measures of the last April should be repealed and the troops withdrawn from Boston. On the 1st of February, Chatham introduced a bill which, could it have passed, would no doubt have averted war, even at the eleventh hour. Besides repealing its vindictive measures, Parliament was to renounce forever the right of taxing the colonies, while retaining the right of regulating the commerce of the whole empire; and the Americans were to defray the expenses of their own governments by taxes voted in their colonial assemblies. A few weeks later, in the House of Commons, Burke argued that the abstract right of Parliament to tax the colonies was not worth contending for, and he urged that on large grounds of expediency it should be

Debates in Parliament.

abandoned, and that the vindictive acts should be repealed. But both Houses, by large majorities, refused to adopt any measures of conciliation, and in a solemn joint address to the king declared themselves ready to support him to the end in the policy upon which he had entered. Massachusetts was declared to be in a state of rebellion, and acts were passed closing all the ports of New England, and prohibiting its fishermen from access to the Newfoundland fisheries. At the same time it was voted to increase the army at Boston to 10,000 men, and to supersede Gage, who had in all these months accomplished so little with his four regiments. As people in England had utterly failed to comprehend the magnitude of the task assigned to Gage, it was not strange that they should seek to account for his inaction by doubting his zeal and ability. No less a person than David Hume saw fit to speak of him as a "lukewarm coward." William Howe, member of Parliament for the liberal constituency of Nottingham, was chosen to supersede him. In his speeches as candidate for election only four months ago, Howe had declared himself opposed to the king's policy, had asserted that no army that England could raise would be able to subdue the Americans, and, in reply to a question, had promised that if offered a command in America he would refuse it. When he now consented to take Gage's place as commander-in-chief, the people of Nottingham scolded him roundly for breaking his word.

William Howe.

It would be unfair, however, to charge Howe

with conscious breach of faith in this matter. His appointment was itself a curious symptom of the element of vacillation that was apparent in the whole conduct of the ministry, even when its attitude professed to be most obstinate and determined. With all his obstinacy, the king did not really wish for war, — much less did Lord North; and the reason for Howe's appointment was simply that he was the brother to the Lord Howe who had fallen at Ticonderoga, and whose memory was idolized by the men of New England. Lord North announced that, in dealing with his misguided American brethren, his policy would be always to send the olive branch in company with the sword; and no doubt Howe really felt that, by accepting a command offered in such a spirit, he might more efficiently serve the interests of humanity and justice than by leaving it open for some one of cruel and despotic temper, whose zeal might outrun even the wishes of the obdurate king. At the same time, his brother Richard, Lord Howe, a seaman of great ability, was appointed admiral of the fleet for America, and was expressly entrusted with the power of offering terms to the colonies. Sir Henry Clinton and John Burgoyne, both of them in sympathy with the king's policy, were appointed to accompany Howe as lieutenant-generals.

Richard, Lord Howe.

The conduct of the ministry, during this most critical and trying time, showed great uneasiness. When leave was asked for Franklin to present the case for the Continental Congress, and to defend it before the House of Commons, it was refused. Yet all through the winter the ministry were continually appealing to Franklin, unofficially and in private, in order to find out how the Americans might be appeased without making any such concessions as would hurt the pride of England. Lord Howe was the most conspicuous agent in these negotiations, which only served to show, over and over again, how the main root of the trouble was the incapacity of the British official mind to understand the character of the American people and the new political situation created by the enormous growth of the colonies. How to conciliate the Americans without giving up a single one of the false positions which the king had taken was the problem, and no wonder that Franklin soon perceived it to be insolvable, and made up his mind to go home.

Franklin returns to America.

He had now stayed in England for several years, as agent for Pennsylvania and for Massachusetts. He had shown himself a consummate diplomatist, of that rare school which deceives by telling unwelcome truths, and he had some unpleasant encounters with the king and the king's friends. Now in March, 1775, seeing clearly that he could be of no further use in averting an armed struggle, he returned to America. Franklin's return was not, in form, like that customary withdrawal of an ambassador which heralds and proclaims a state of war. But practically it was the snapping of the last diplomatic link between the colonies and the mother-country.

Still the ministry, with all its uneasiness, did not believe that war was close at hand. It was thought that the middle colonies, and especially

New York, might be persuaded to support the government, and that New England, thus isolated, would not venture upon armed resistance to the overwhelming power of Great Britain. The hope was not wholly unreasonable; for the great middle colonies, though conspicuous for material prosperity, were somewhat lacking in force of political ideas. In New York and Pennsylvania the non-English population was relatively far more considerable than in the Southern or the New England colonies. A considerable proportion of the population had come from the continent of Europe, and the principles of constitutional government were not so thoroughly inwrought into the innermost minds and hearts of the people, the pulse of liberty did not beat so quickly here, as in the purely English commonwealths of Virginia and Massachusetts. In Pennsylvania and New Jersey the Quakers were naturally opposed to a course of action that must end in war; and these very honourable motives certainly contributed to weaken the resistance of these colonies to the measures of the government. In New York there were further special reasons for the existence of a strong loyalist feeling. The city of New York had for many years been the headquarters of the army and the seat of the principal royal government in America. It was not a town, like Boston, governing itself in town-meeting, but its municipal affairs were administered by a mayor, appointed by the king. Unlike Boston and Philadelphia, the interests of the city of New York were almost purely commercial, and there was nothing

The middle colonies.

to prevent the little court circle there from giving the tone to public opinion. The Episcopal Church, too, was in the ascendant, and there was a not unreasonable prejudice against the Puritans of New England for their grim intolerance of Episcopalians and their alleged antipathy to Dutchmen. The province of New York, moreover, had a standing dispute with its eastern neighbours over the ownership of the Green Mountain region. This beautiful country had been settled by New England men, under grants from the royal governors of New Hampshire; but it was claimed by the people of New York, and the controversy sometimes waxed hot and gave rise to very hard feelings. Under these circumstances, the labours of the ministry to secure this central colony seemed at times likely to be crowned with success. The assembly of New York refused to adopt the non-importation policy enjoined by the Continental Congress, it refused to print letters of the committee of correspondence, and it refused to choose delegates to the second Congress which was to be held in May. The ministry, in return, sought to corrupt New York by exempting it from the commercial restrictions placed upon the neighbouring colonies, and by promising to confirm its alleged title to the territory of Vermont. All these hopes proved fallacious, however. In spite of appearances, the majority of the people of New York were thoroughly patriotic, and needed only an opportunity for organization. In April, under the powerful leadership of Philip Schuyler and the Livingstons, a convention was held, delegates were

Lord North's mistaken hopes of securing New York.

chosen to attend the Congress, and New York fell into line with the other colonies. As for Pennsylvania, in spite of its peaceful and moderate temper, it had never shown any signs of willingness to detach itself from the nascent union.

News travelled with slow pace in those days, and as late as the middle of May, Lord North, confident of the success of his schemes in New York, and unable to believe that the yeomanry of Massachusetts would fight against regular troops, declared cheerfully that this American business was not so alarming as it seemed, and everything would no doubt be speedily settled without bloodshed!

Affairs in Massachusetts.

Great events had meanwhile happened in Massachusetts. All through the winter the resistance to General Gage had been passive, for the lesson had been thoroughly impressed upon the mind of every man, woman, and child in the province that, in order to make sure of the entire sympathy of the other colonies, Great Britain must be allowed to fire the first shot. The Regulating Act had none the less been silently defied, and neither councillors nor judges, neither sheriffs nor jurymen, could be found to serve under the royal commission. It is striking proof of the high state of civilization attained by this commonwealth that although for nine months the ordinary functions of government had been suspended, yet the affairs of every-day life had gone on without friction or disturbance. Not a drop of blood had been shed, nor had any one's property been injured. The companies of yeomen meeting at eventide to drill on the village green, and now and then the cart laden with powder and ball that dragged slowly over the steep roads on its way to Concord, were the only outward signs of an unwonted state of things. Not so, however, in Boston. There the blockade of the harbour had wrought great hardship for the poorer people. Business was seriously interfered with, many persons were thrown out of employment, and in spite of the generous promptness with which provisions had been poured in from all parts of the country, there was great suffering through scarcity of fuel and food. Still there was but little complaint and no disorder. The leaders were as resolute as ever, and the people were as resolute as their leaders. As the 5th of March drew near, several British officers were heard to declare that any one who should dare to address the people in the Old South Church on this occasion would surely lose his life. As soon as he heard of these threats, Joseph Warren solicited for himself the dangerous honour, and at the usual hour delivered a stirring oration upon "the baleful influence of standing armies in time of peace." The concourse in the church was so great that when the orator arrived every approach to the pulpit was blocked up; and rather than elbow his way through the crowd, which might lead to some disturbance, he procured a ladder, and climbed in through a large window at the back of the pulpit. About forty British officers were present, some of whom sat on the pulpit steps, and sought to annoy the speaker with groans and hisses, but everything passed off quietly.

Warren's oration at the Old South.

The boldness of Adams and Hancock in attending this meeting was hardly less admirable than that of Warren in delivering the address. It was no secret that Gage had been instructed to watch his opportunity to arrest Samuel Adams and "his willing and ready tool," that "terrible desperado," John Hancock, and send them over to England to be tried for treason. Here was an excellent opportunity for seizing all the patriot leaders at once; and the meeting itself, moreover, was a town-meeting, such as Gage had come to Boston expressly to put down. Nothing more calmly defiant can be imagined than the conduct of people and leaders under these circumstances. But Gage had long since learned the temper of the people so well that he was afraid to proceed too violently.

Attempt to corrupt Samuel Adams

At first he had tried to corrupt Samuel Adams with offers of place or pelf; but he found, as Hutchinson had already declared, that such was "the obstinate and inflexible disposition of this man that he never would be conciliated by any office or gift whatsoever." The dissolution of the assembly, of which Adams was clerk, had put a stop to his salary, and he had so little property laid by as hardly to be able to buy bread for his family. Under these circumstances, it occurred to Gage that perhaps a judicious mixture of threat with persuasion might prove effectual. So he sent Colonel Fenton with a confidential message to Adams. The officer, with great politeness, began by saying that "an adjustment of the existing disputes was very desirable; that he was authorized by Governor Gage to assure him that he had been empowered to confer upon him such benefits as would be satisfactory, upon the condition that he would engage to cease in his opposition to the measures of government, and that it was the advice of Governor Gage to him not to incur the further displeasure of his Majesty; that his conduct had been such as made him liable to the penalties of an act of Henry VIII., by which persons could be sent to England for trial, and, by changing his course, he would not only receive great personal advantages, but would thereby make his peace with the king." Adams listened with apparent interest to this recital until the messenger had concluded. Then rising, he replied, glowing with indignation: "Sir, I trust I have long since made my peace with the King of kings. No personal consideration shall induce me to abandon the righteous cause of my country. Tell Governor Gage it is the advice of Samuel Adams to him no longer to insult the feelings of an exasperated people."

Orders to arrest Adams and Hancock.

Toward the end of the winter Gage received peremptory orders to arrest Adams and Hancock, and send them to England for trial. One of the London papers gayly observed that in all probability Temple Bar "will soon be decorated with some of the patriotic noddles of the Boston saints." The provincial congress met at Concord on the 22d of March, and after its adjournment, on the 15th of April, Adams and Hancock stayed a few days at Lexington, at the house of their friend, the Rev. Jonas Clark. It would doubtless be easier to seize them there than in Boston, and, accordingly, on the night of the 18th Gage dispatched a force of 800 troops, under Lieutenant-Colonel Smith, to march to Lexington, and, after seizing the patriot leaders, to proceed to Concord, and capture or destroy the military stores which had for some time been collecting there. At ten in the evening the troops were rowed across Charles river, and proceeded by a difficult and unfrequented route through the marshes of East Cambridge, until, after four miles, they struck into the highroad for Lexington. The greatest possible secrecy was observed, and stringent orders were given that no one should be allowed to leave Boston that night. But Warren divined the purpose of the movement, and sent out Paul Revere by way of Charlestown, and William Dawes by way of Roxbury, to give the alarm. At that time there was no bridge across Charles river lower than the one which now connects Cambridge with Allston. Crossing the broad river in a little boat, under the very guns of the Somerset man-of-war, and waiting on the farther bank until he learned, from a lantern suspended in the belfry of the North Church, which way the troops had gone, Revere took horse and galloped over the Medford road to Lexington, shouting the news at the door of every house that he passed. Reaching Mr. Clark's a little after midnight, he found the house guarded by eight minute-men, and the sergeant warned him not to make a noise and disturb the inmates. "Noise!" cried Revere. "You 'll soon have noise enough; the regulars are coming!" Hancock, recognizing the voice, threw up the window, and ordered the

Paul Revere's ride.

guard to let him in. On learning the news, Hancock's first impulse was to stay and take command of the militia; but it was presently agreed that there was no good reason for his doing so, and shortly before daybreak, in company with Adams, he left the village.

Meanwhile, the troops were marching along the main road; but swift and silent as was their advance, frequent alarm-bells and signal-guns, and lights twinkling on distant hill-tops, showed but too plainly that the secret was out. Colonel Smith then sent Major Pitcairn forward with six companies of light infantry to make all possible haste in securing the bridges over Concord river, while at the same time he prudently sent back to Boston for reinforcements. When Pitcairn reached Lexington, just as the rising sun was casting long shadows across the village green, he found himself confronted by some fifty minute-men under command of Captain John Parker, — grandfather of Theodore Parker, — a hardy veteran, who, fifteen years before, had climbed the heights of Abraham by the side of Wolfe. "Don't fire unless you are fired on," said Parker; "but if they want a war, it may as well begin here." "Disperse, ye villains!" shouted Pitcairn. "Damn you, why don't you disperse?" And as they stood motionless he gave the order to fire. As the soldiers hesitated to obey, he discharged his own pistol and repeated the order, whereupon a deadly volley slew eight of the minute-men and wounded ten. At this moment the head of Smith's own column seemed to have come into sight, far

Pitcairn fires upon the yeomanry, April 19, 1775.

down the road. The minute-men had begun to return the fire, when Parker, seeing the folly of resistance, ordered them to retire. While this was going on, Adams and Hancock were walking across the fields toward Woburn; and as the crackle of distant musketry reached their ears, the eager Adams;— his soul aglow with the prophecy of the coming deliverance of his country — exclaimed, "Oh, what a glorious morning is this!" From Woburn the two friends went on their way to Philadelphia, where the second Continental Congress was about to assemble.

Some precious minutes had been lost by the British at Lexington, and it soon became clear that the day was to be one in which minutes could ill be spared. By the time they reached Concord, about seven o'clock, the greater part of the stores had been effectually hidden, and minute-men were rapidly gathering from all quarters. After posting small forces to guard the bridges, the troops set fire to the court-house, cut down the liberty-pole, disabled a few cannon, staved in a few barrels of flour, and hunted unsuccessfully for arms and ammunition, until an unexpected incident put a stop to their proceedings. When the force of minute-men, watching events from the hill beyond the river, had become increased to more than 400, they suddenly advanced upon the North Bridge, which was held by 200 regulars. After receiving and returning the British fire, the militia, led by Major Buttrick, charged across the narrow bridge, overcame the regulars through weight and numbers, and

The troops repulsed at Concord.

drove them back into the village. They did not follow up the attack, but rested on their arms, wondering, perhaps, at what they had already accomplished, while their numbers were from moment to moment increased by the minute-men from neighbouring villages. A little before noon, though none of the objects of the expedition had been accomplished, Colonel Smith began to realize the danger of his position, and started on his retreat to Boston. His men were in no mood for fight. They had marched eighteen miles, and had eaten little or nothing for fourteen hours. But now, while companies of militia hovered upon both their flanks, every clump of trees and every bit of rising ground by the roadside gave shelter to hostile yeomen, whose aim was true and deadly. Straggling combats ensued from time to time, and the retreating British left nothing undone which brave men could do; but the incessant, galling fire at length threw them into hopeless confusion. Leaving their wounded scattered along the road, they had already passed by the village green of Lexington in disorderly flight, when they were saved by Lord Percy, who had marched out through Brookline and Cambridge to their assistance, with 1,200 men and two field-pieces. Forming his men in a hollow square, Percy inclosed the fugitives, who, in dire exhaustion, threw themselves upon the ground, — "their tongues hanging out of their mouths," says Colonel Stedman, "like those of dogs after a chase." Many had thrown away their muskets, and Pitcairn had lost his horse, with the elegant pistols which fired

Retreating troops rescued by Lord Percy.

the first shots of the War of Independence, and which may be seen to-day, along with other trophies, in the town library of Lexington.

Percy's timely arrival checked the pursuit for an hour, and gave the starved and weary men a chance for food and rest. A few houses were pillaged and set on fire, but at three o'clock General Heath and Dr. Warren arrived on the scene and took command of the militia, and the irregular fight was renewed. When Percy reached Menotomy (now Arlington), seven miles from Boston, his passage was disputed by a fresh force of militia, while pursuers pressed hard on his rear, and it was only after an obstinate fight that he succeeded in forcing his way. The roadside now fairly swarmed with marksmen, insomuch that, as one of the British officers observed, "they seemed to have dropped from the clouds." It became impossible to keep order or to carry away the wounded; and when, at sunset, the troops entered Charlestown, under the welcome shelter of the fleet, it was upon the full run. They were not a moment too soon, for Colonel Pickering, with 700 Essex militia, on the way to intercept them, had already reached Winter Hill; and had their road been blocked by this fresh force they must in all probability have surrendered.

Retreat continued from Lexington to Charlestown.

On this eventful day the British lost 273 of their number, while the Americans lost 93. The expedition had been a failure, the whole British force had barely escaped capture, and it had been shown that the people could not be frightened into submission. It had been shown, too, how efficient the town system of organized militia might prove on a sudden emergency.) The most interesting feature of the day is the rapidity and skill with which the different bodies of minute-men, marching from long distances, were massed at those points on the road where they might most effectually impede the British retreat. The Danvers company marched sixteen miles in four hours to strike Lord Percy at Menotomy. The list of killed and wounded shows that contingents from at least twenty-three towns had joined in the fight before sundown. But though the pursuit was then ended, these men did not return to their homes, but hour by hour their numbers increased. At noon of that day the alarm had reached Worcester. Early next morning, Israel Putnam was ploughing a field at Pomfret, in Connecticut, when the news arrived. Leaving orders for the militia companies to follow, he jumped on his horse, and, riding a hundred miles in eighteen hours, arrived in Cambridge on the morning of the 21st, just in time to meet John Stark with the first company from New Hampshire. At midday of the 20th the college green at New Haven swarmed with eager students and citizens, and Captain Benedict Arnold, gathering sixty volunteers from among them, placed himself at their head and marched for Cambridge, picking up recruits and allies at all the villages on the way. And thus, from every hill and valley in New England, on they came, till, by Saturday night, Gage found himself besieged in Boston by a rustic army of 16,000 men.

Rising of the country; the British besieged in Boston.

When the news of this affair reached England, five weeks later, it was received at first with incredulity, then with astonishment and regret. Slight as the contest had been, it remained undeniable that British troops had been defeated by what in England was regarded as a crowd of "peasants;" and it was felt besides that the chances for conciliation had now been seriously diminished. Burke said that now that the Americans had once gone so far as this, they could hardly help going farther; and in spite of the condemnation that had been lavished upon Gage for his inactivity, many people were now inclined to find fault with him for having precipitated a conflict just at the time when it was hoped that, with the aid of the New York loyalists, some sort of accommodation might be effected. There is no doubt that the news from Lexington thoroughly disconcerted the loyalists of New York for the moment, and greatly strengthened the popular party there. In a manifesto addressed to the city of London, the New York committee of correspondence deplored the conduct of Gage as rash and violent, and declared that all the horrors of civil war would never bring the Americans to submit to the unjust acts of Parliament. When Hancock and Adams arrived, on their way to the Congress, they were escorted through the city with triumphal honours. In Pennsylvania steps were immediately taken for the enlistment and training of a colonial militia, and every colony to the south of it followed the example. The Scotch-Irish patriots of Mecklenburg county, in North Caro-

Effects of the news.

lina, ventured upon a measure more decided than any that had yet been taken in any part of the country. On May 31st, the county committee of Mecklenburg affirmed that the joint address of the two Houses of Parliament to the king, in February, had virtually "annulled and vacated all civil and military commissions granted by the Crown, and suspended the constitutions of the colonies;" and that consequently "the provincial congress of each province, under the direction of the great Continental Congress, is invested with all the legislative and executive powers within their respective provinces, and that no other legislative or executive power does or can exist at this time in any of these colonies." In accordance with this state of things, rules were adopted "for the choice of county officers, to exercise authority by virtue of this choice and independently of the British Crown, until Parliament should resign its arbitrary pretensions." These bold resolves were entrusted to the North Carolina delegates to the Continental Congress, but were not formally brought before that body, as the delegates thought it best to wait for a while longer the course of events.

Mecklenburg County Resolves, May 31, 1775.

Some twenty years later they gave rise to the legend of the Mecklenburg Declaration of Independence. The early writers of United States history passed over the proceedings of May 31st in silence, and presently the North Carolina patriots tried to supply an account of them from memory. Their traditional account was not published until 1819, when it was

Legend of the Mecklenburg "Declaration of Independence."

found to contain a spurious document, giving the substance of some of the foregoing resolves, decorated with phrases borrowed from the great Declaration of Independence of 1776. This document purported to have been drawn up and signed at a county meeting on the 20th of May. A fierce controversy sprang up over the genuineness of the document, which was promptly called in question. For a long time many people believed in it, and were inclined to charge Jefferson with having plagiarized from it in writing the Declaration of Independence. But a minute investigation of all the newspapers of May, 1775, has shown that no such meeting was held on the 20th, and that no such document was made public. The story of the Mecklenburg Declaration is simply a legend based upon the distorted recollection of the real proceedings of May 31st.

Meanwhile, in New England, the warlike feeling had become too strong to be contented merely with defensive measures. No sooner had Benedict Arnold reached Cambridge than he suggested to Dr. Warren that an expedition ought to be sent without delay to capture Ticonderoga and Crown Point. These fortresses commanded the northern approaches to the Hudson river, the strategic centre of the whole country, and would be of supreme importance either in preparing an invasion of Canada or in warding off an invasion of New York. Besides this, they contained a vast quantity of military stores, of which the newly gathered army stood in sore need. The idea found favour at once. Arnold received a colonel's

Benedict Arnold and Ethan Allen.

commission from the Massachusetts Congress, and was instructed to raise 400 men among the Berkshire Hills, capture the fortresses, and superintend the transfer of part of their armament to Cambridge. When Arnold reached the wild hillsides of the Hoosac range, he found that he had a rival in the enterprise. The capture of Ticonderoga had also been secretly planned in Connecticut, and was entrusted to Ethan Allen, the eccentric but sagacious author of that now-forgotten deistical book, "The Oracles of Reason." Allen was a leading spirit among the "Green Mountain Boys," an association of Vermont settlers formed for the purpose of resisting the jurisdiction of New York, and his personal popularity was great. On the 9th of May Arnold overtook Allen and his men on their march toward Lake Champlain, and claimed the command of the expedition on the strength of his commission from Massachusetts; but the Green Mountain Boys were acting partly on their own account, partly under the direction of Connecticut. They cared nothing for the authority of Massachusetts, and knew nothing of Arnold; they had come out to fight under their own trusted leader. But few of Arnold's own men had as yet assembled, and his commission could not give him command of Vermonters, so he joined the expedition as a volunteer. On reaching the lake that night, they found there were not nearly enough row-boats to convey the men across. But delay was not to be thought of. The garrison must not be put on its guard. Accordingly, with only eighty-three men, Allen and Arnold crossed the lake at daybreak of the 10th, and entered Ticonderoga side by side. The little garrison, less than half as many in number, as it turned out, was completely surprised, and the stronghold was taken without a blow. As the commandant jumped out of bed, half awake, he confusedly inquired of Allen by whose authority he was acting. "In the name of the Great Jehovah and the Continental Congress!" roared the bellicose philosopher, and the commandant, seeing the fort already taken, was fain to acquiesce. At the same time Crown Point surrendered to another famous Green Mountain Boy, Seth Warner, and thus more than two hundred cannon, with a large supply of powder and ball, were obtained for the New England army. A few days later, as some of Arnold's own men arrived from Berkshire, he sailed down Lake Champlain, and captured St. John's with its garrison; but the British recovered it in the course of the summer, and planted such a force there that in the autumn we shall see it able to sustain a siege of fifty days.

Capture of Ticonderoga and Crown Point, May 10, 1775.

Neither Connecticut nor Massachusetts had any authority over these posts save through right of conquest. As it was Connecticut that had set Allen's expedition on foot, Massachusetts yielded the point as to the disposal of the fortresses and their garrisons. Dr. Warren urged the Connecticut government to appoint Arnold to the command, so that his commission might be held of both colonies; but Connecticut preferred to retain Allen, and in July Arnold returned to Cambridge to mature his remarkable plan for invading Canada through the trackless wilderness of Maine. His slight disagreement with Allen bore evil fruit. As is often the case in such affairs, the men were more zealous than their commanders; there were those who denounced Arnold as an interloper, and he was destined to hear from them again and again.

Second meeting of the Continental Congress, May 10, 1775.

On the same day on which Ticonderoga surrendered, the Continental Congress met at Philadelphia. The Adamses and the Livingstons, Jay, Henry, Washington, and Lee were there, as also Franklin, just back from his long service in England. Of all the number, John Adams and Franklin had now, probably, come to agree with Samuel Adams that a political separation from Great Britain was inevitable; but all were fully agreed that any consideration of such a question was at present premature and uncalled for. The Congress was a body which wielded no technical legal authority; it was but a group of committees, assembled for the purpose of advising with each other regarding the public weal. Yet something very like a state of war existed in a part of the country, under conditions which intimately concerned the whole, and in the absence of any formally constituted government something must be done to provide for such a crisis. The spirit of the assembly was well shown in its choice of a president. Peyton Randolph being called back to Virginia to preside over the colonial assembly, Thomas Jefferson was sent to the Congress in his stead; and it also became necessary for Congress to choose a president to succeed him. The proscribed John Hancock was at once chosen, and Benjamin Harrison, in conducting him to the chair, said, "We will show Great Britain how much we value her proscriptions." To the garrisoning of Ticonderoga and Crown Point by Connecticut, the Congress consented only after much hesitation, since the capture of these posts had been an act of offensive warfare. But without any serious opposition, in the name of the "United Colonies," the Congress adopted the army of New England men besieging Boston as the "Continental Army," and proceeded to appoint a commander-in-chief to direct its operations. Practically, this was the most important step taken in the whole course of the War of Independence, and the wisdom shown in the appointment was consummate. Nothing less, indeed, than the whole issue of the struggle, for ultimate defeat or for ultimate victory, turned upon the selection to be made at this crisis. For nothing can be clearer than that in any other hands than those of George Washington the military result of the war must have been speedily disastrous to the Americans. In appointing a Virginian to the command of a New England army, the Congress showed rare wisdom. It would well have accorded with local prejudices had a New England general been appointed. John Hancock greatly desired the appointment, and seems to have been chagrined at not receiving it. But it was wisely decided that the common interest of all Americans could in no way be more thoroughly engaged in the war than by putting

Appointment of Washington to command the Continental army.

the New England army in charge of a general who represented in his own person the greatest of the Southern colonies. Washington was now commander of the local militia of Virginia, and sat in Congress in his colonel's uniform. His services in saving the remnant of Braddock's ill-fated army, and afterwards in the capture of Fort Duquesne, had won for him a military reputation greater than that of any other American. Besides this, there was that which, from his early youth, had made it seem right to entrust him with commissions of extraordinary importance. Nothing in Washington's whole career is more remarkable than the fact that when a mere boy of twenty-one he should have been selected by the governor of Virginia to take charge of that most delicate and dangerous diplomatic mission to the Indian chiefs and the French commander at Venango. Consummate knowledge of human nature as well as of wood-craft, a courage that no threats could daunt and a clear intelligence that no treachery could hoodwink, were the qualities absolutely demanded by such an undertaking; yet the young man acquitted himself of his perilous task not merely with credit, but with splendour. As regards book-lore, his education had been but meagre, yet he possessed in the very highest degree the rare faculty of always discerning the essential facts in every case, and interpreting them correctly. In the Continental Congress there sat many who were superior to him in learning and eloquence; but "if," said Patrick Henry, "you speak of solid information and sound judgment, Colonel Washington is unquestionably the greatest man upon that floor." Thus did that wonderful balance of mind — so great that in his whole career it would be hard to point out a single mistake — already impress his ablest contemporaries. Hand in hand with this rare soundness of judgment there went a completeness of moral self-control, which was all the more impressive inasmuch as Washington's was by no means a tame or commonplace nature, such as ordinary power of will would suffice to guide. He was a man of intense and fiery passions. His anger, when once aroused, had in it something so terrible that strong men were cowed by it like frightened children. This prodigious animal nature was habitually curbed by a will of iron, and held in the service of a sweet and tender soul, into which no mean or unworthy thought had ever entered. Whole-souled devotion to public duty, an incorruptible integrity which no appeal to ambition or vanity could for a moment solicit, — these were attributes of Washington, as well marked as his clearness of mind and his strength of purpose. And it was in no unworthy temple that Nature had enshrined this great spirit. His lofty stature (exceeding six feet), his grave and handsome face, his noble bearing and courtly grace of manner, all proclaimed in Washington a king of men.

The choice of Washington for commander-in-chief was suggested and strongly urged by John Adams, and when, on the 15th of June, the nomination was formally made by Thomas Johnson of Maryland, it was unanimously confirmed. Then

Washington, rising, said with great earnestness: "Since the Congress desire, I will enter upon the momentous duty, and exert every power I possess in their service and for the support of the glorious cause. But I beg it may be remembered by every gentleman in the room that I this day declare, with the utmost sincerity, I do not think myself equal to the command I am honoured with." He refused to take any pay for his services, but said he would keep an accurate account of his personal expenses, which Congress might reimburse, should it see fit, after the close of the war.

While these things were going on at Philadelphia, the army of New England men about Boston was busily pressing, to the best of its limited ability, the siege of that town. The army extended in a great semicircle of sixteen miles, — averaging about a thousand men to the mile, — all the way from Jamaica Plain to Charlestown Neck. The headquarters were at Cambridge, where some of the university buildings were used for barracks, and the chief command had been entrusted to General Artemas Ward, under the direction of the committee of safety. Dr. Warren had succeeded Hancock as president of the provincial congress, which was in session at Watertown. The army was excellent in spirit, but poorly equipped and extremely deficient in discipline. Its military object was to compel the British troops to evacuate Boston and take to their ships; for as there was no American fleet, anything like the destruction or capture of the British force was manifestly impossible. The only way in which Boston

Siege of Boston.

could be made untenable for the British was by seizing and fortifying some of the neighbouring hills which commanded the town, of which the most important were those in Charlestown on the north and in Dorchester on the southeast. To secure these hills was indispensable to Gage, if he was to keep his foothold in Boston; and as soon as Howe, Clinton, and Burgoyne arrived, on the 25th of May, with reinforcements which raised the British force to 10,000 men, a plan was laid for extending the lines so as to cover both Charlestown and Dorchester. Feeling now confident of victory, Gage issued a proclamation on June 12th, offering free pardon to all rebels who should lay down their arms and return to their allegiance, saving only those ringleaders, John Hancock and Samuel Adams, whose crimes had been "too flagitious to be condoned." At the same time, all who should be taken in arms were threatened with the gallows. In reply to this manifesto, the committee of safety, having received intelligence of Gage's scheme, ordered out a force of 1,200 men, to forestall the governor, and take possession of Bunker Hill in Charlestown. At sunset of the 16th this brigade was paraded on Cambridge Common, and after prayer had been offered by Dr. Langdon, president of the university, they set out on their enterprise, under command of Colonel Prescott of Pepperell, a veteran of the French war, grandfather of one of the most eminent of American historians. On reaching the grounds, a consultation was held, and it was decided, in accordance with the general purpose,

Gage's proclamation.

if not in strict conformity to the letter of the order, to push on farther and fortify the eminence known as Breed's Hill, which was connected by a ridge with Bunker Hill, and might be regarded as part of the same locality.

Americans occupy Bunker Hill.

The position of Breed's Hill was admirably fitted for annoying the town and the ships in the harbour, and it was believed that, should the Americans succeed in planting batteries there, the British would be obliged to retire from Boston. There can be little doubt, however, that in thus departing from the strict letter of his orders Prescott made a mistake, which might have proved fatal, had not the enemy blundered still more seriously. The advanced position on Breed's Hill was not only exposed to attacks in the rear from an enemy who commanded the water, but the line of retreat was ill secured, and, by seizing upon Charlestown Neck, it would have been easy for the British, with little or no loss, to have compelled Prescott to surrender. From such a disaster the Americans were saved by the stupid contempt which the enemy felt for them.

Reaching Breed's Hill about midnight, Colonel Prescott's men began throwing up intrenchments. At daybreak they were discovered by the sailors in the harbour, and a lively cannonade was kept up through the forenoon by the enemy's ships; but it produced little effect, and the strength of the American works increased visibly hour by hour. It was a beautiful summer day, bathed in brightest sunshine, and through the clear dry air every movement of the spadesmen on the hill-top and the sailors on their decks could be distinctly seen from a great distance. The roar of the cannon had called out everybody, far and near, to see what was going on, and the windows and house-tops in Boston were crowded with anxious spectators. During the night General Putnam had come upon the scene, and turned his attention to fortifying the crest of Bunker Hill, in order to secure the line of retreat across Charlestown Neck. In the course of the forenoon Colonel Stark arrived with reinforcements, which were posted behind the rail fence on the extreme left, to ward off any attempt of the British to turn their flank by a direct attack. At the same time, Dr. Warren, now chief executive officer of Massachusetts, and just appointed major-general, hastened to the battle-field; replying to the prudent and affectionate remonstrance of his friend Elbridge Gerry, "Dulce et decorum est pro patria mori." Arriving at the redoubt, he refused the command expressly tendered him, saying that he should be only too glad to serve as volunteer aid, and learn his first lesson under so well tried a soldier as Prescott. This modest heroism was typical of that memorable day, to the events of which one may well apply the Frenchman's dictum, "C'est magnifique, mais ce n'est pas la guerre!" A glorious day it was in history, but characterized, on both the British and the American sides, by heroism rather than by military skill or prudence.

Arrival of Putnam, Stark, and Warren, June 17, 1775.

During the forenoon Gage was earnestly discussing with the three new generals the best means of ousting the Americans from their position on Breed's Hill. There was one sure and obvious method, — to go around by sea and take possession of Charlestown Neck, thereby cutting off the Americans from the mainland and starving them out. But it was thought that time was too precious to admit of so slow a method. Should the Americans succeed, in the course of the afternoon, in planting a battery of siege guns on Breed's Hill, the British position in Boston would be endangered. A direct assault was preferred, as likely to be more speedily effective. It was unanimously agreed that these "peasants" could not withstand the charge of 3,000 veteran soldiers, and it was gravely doubted if they would stay and fight at all. Gage accordingly watched the proceedings, buoyant with hope. In a few hours the disgrace of Lexington would be wiped out, and this wicked rebellion would be ended. At noonday the troops began crossing the river in boats, and at three o'clock they prepared to storm the intrenchments. They advanced in two parties, General Howe toward the rail-fence, and General Pigott toward the redoubt, and the same fate awaited both. The Americans reserved fire until the enemy had come within fifty yards, when all at once they poured forth such a deadly volley that the whole front rank of the British was mowed as if by the sudden sweep of a scythe. For a few minutes the gallant veterans held their ground and returned the fire; but presently an indescribable shudder ran through the line, and

Gage decides to try an assault.

First assault repulsed.

they gave way and retreated down the hillside in disorder, while the Americans raised an exultant shout, and were with difficulty restrained by their officers from leaping over the breastworks and pursuing.

A pause now ensued, during which the village of Charlestown was set on fire by shells from the fleet, and soon its four hundred wooden houses were in a roaring blaze, while charred timbers strewed the lawns and flower-beds, and the sky was blackened with huge clouds of smoke. If the purpose of this wholesale destruction of property was, as some have thought, to screen the second British advance, the object was not attained, for a light breeze drove the smoke the wrong way. As the bright red coats, such excellent targets for trained marksmen, were seen the second time coming up the slope, the Americans, now cool and confident, withheld their fire until the distance was less than thirty yards. Then, with a quick succession of murderous discharges, such havoc was wrought in the British lines as soon to prove unendurable. After a short but obstinate struggle the lines were broken, and the gallant troops retreated hastily, leaving the hillside covered with their dead and wounded. All this time the Americans, in their sheltered position, had suffered but little.

Second assault repulsed.

So long a time now elapsed that many persons began to doubt if the British would renew the assault. Had the organization of the American army been better, such reinforcements of men and ammunition might by this time have arrived from

Cambridge that any further attack upon the hill would be sure to prove fruitless. But all was confusion at headquarters. General Ward was ill furnished with staff officers, and wrong information was brought, while orders were misunderstood. And besides, in his ignorance of the extent of Gage's plans, General Ward was nervously afraid of weakening his centre at Cambridge. Three regiments were sent over too late to be of any use, and meanwhile Prescott, to his dismay, found that his stock of powder was nearly exhausted. While he was making ready for a hand-to-hand fight, the British officers were holding a council of war, and many declared that to renew the attack would be simply useless butchery. On the other hand, General Howe observed, "To be forced to give up Boston would, gentlemen, be very disagreeable to us all." The case was not really so desperate as this, for the alternative of an attack upon Charlestown Neck still remained open, and every consideration of sound generalship now prescribed that it should be tried. But Howe could not bear to acknowledge the defeat of his attempts to storm, and accordingly, at five o'clock, with genuine British persistency, a third attack was ordered. For a moment the advancing columns were again shaken by the American fire, but the last cartridges were soon spent, and by resolute bayonet charges and irregular volleys that could not be returned the Americans were slowly driven from their works and forced to retreat over Charlestown Neck, while the whole disputed

Prescott's powder gives out.

Third assault succeeds; the British take the hill.

ground, including the summit of Bunker Hill, passed into the hands of the British.

In this battle, in which not more than one hour was spent in actual fighting, the British loss in killed and wounded was 1,054, or more than one third of the whole force engaged, including an unusually large proportion of officers. The American loss, mainly incurred at the rail-fence and during the final hand-to-hand struggle at the redoubt, was 449, probably about one fourth of the whole force engaged.

British and American losses.

On the British side, one company of grenadiers came out of the battle with only five of its number left unhurt. Every officer on General Howe's staff was cut down, and only one survived his wounds. The gallant Pitcairn, who had fired the first shot of the war, fell while entering the redoubt, and a few moments later the Americans met with an irreparable loss in the death of General Warren, who was shot in the forehead as he lingered with rash obstinacy on the scene, loath to join in the inevitable retreat. Another volunteer aid, not less illustrious than Warren, fought on Bunker Hill that day, and came away scatheless. Since the brutal beating which he had received at the coffee-house nearly six years before, the great intellect of James Otis had suffered well-nigh total wreck. He was living, harmlessly insane, at the house of his sister, Mercy Warren, at Watertown, when he witnessed the excitement and listened to the rumour of battle on the morning of the 17th of June. With touching eagerness to strike a blow for the cause in which he had already suffered so dreadful a martyrdom, Otis stole away from home, borrowed a musket at some roadside farmhouse, and hastened to the battlefield, where he fought manfully, and after all was over made his way home, weary and faint, a little before midnight.

Excessive slaughter; significance of the battle.

Though small in its dimensions, if compared with great European battles, or with the giant contests of our own civil war, the struggle at Bunker Hill is memorable and instructive, even from a purely military point of view. Considering the numbers engaged and the short duration of the fight, the destruction of life was enormous. Of all the hardest-fought fields of modern times, there have been very few indeed in which the number of killed and wounded has exceeded one fourth of the whole force engaged. In its bloodiness and in the physical conditions of the struggle, the battle of Bunker Hill resembles in miniature the tremendous battles of Fredericksburg and Cold Harbor. To ascend a rising ground and storm well-manned intrenchments has in all ages been a difficult task; at the present day, with the range and precision of our modern weapons, it has come to be almost impossible. It has become a maxim of modern warfare that only the most extraordinary necessity can justify a commander in resorting to so desperate a measure. He must manœuvre against such positions, cut them off by the rear, or deprive them of their value by some flanking march; but he must not, save as a forlorn hope, waste precious human lives in an effort to storm them that is almost sure to prove fruitless. For our means of destroying life have become so powerful and so accurate that, when skilfully wielded from commanding positions, no human gallantry can hope to withstand them. As civilization advances, warfare becomes less and less a question of mere personal bravery, and more and more a question of the application of resistless physical forces at the proper points; that is to say, it becomes more and more a purely scientific problem of dynamics. Now at Bunker Hill, though the Americans had not our modern weapons of precision, yet a similar effect was wrought by the remarkable accuracy of their aim, due to the fact that they were all trained marksmen, who waited coolly till they could fire at short range, and then wasted no shots in random firing. Most of the British soldiers who fell in the two disastrous charges of that day were doubtless picked off as partridges are picked off by old sportsmen, and thus is explained the unprecedented slaughter of officers. Probably nothing quite like this had yet been seen in the history of war, though the principle had been similar in those wonderful trials of the long-bow in such mediæval battles as Crécy and Dupplin Moor. Against such odds even British pluck and endurance could not possibly prevail. Under these circumstances, had the Americans been properly supplied with powder, Howe could no more have taken Bunker Hill by storm than Burnside could take the heights of Fredericksburg.

The moral effect of the battle of Bunker Hill, both in America and Europe, was remarkable. It was for the British a decided and important victory, inasmuch as they not only gained the ground for which the battle was fought, but by so doing they succeeded in keeping their hold upon Boston for nine months longer.

Its moral effect.

Nevertheless, the moral advantage was felt to be entirely on the side of the Americans. It was they who were elated by the day's work, while it was the British who were dispirited. The belief that Americans could not fight was that day dispelled forever. British officers who remembered Fontenoy and Minden declared that the firing at Bunker Hill was the hottest they had ever known, and, with an exaggeration which was pardonable as a reaction from their former ill-judged contempt, it was asserted that the regulars of France were less formidable foes than the militia of New England. It was keenly felt that if a conquest of a single strategic position had encountered such stubborn resistance, the task of subjugating the United Colonies was likely to prove a hard one. "I wish we could sell them another hill at the same price," said General Greene. Vergennes, the French minister of foreign affairs, exclaimed that with two more such victories England would have no army left in America. Washington said there could now be no doubt that the liberties of the people were secure. While Franklin, taking extreme ground, declared that England had lost her colonies forever.

CHAPTER IV.

INDEPENDENCE.

On the 2d of July, 1775, after a journey of eleven days, General Washington arrived in Boston from Philadelphia, and on the following day, under the shade of the great elm-tree which still stands hard by Cambridge Common, he took command of the Continental army, which as yet was composed entirely of New Englanders. Of the 16,000 men engaged in the siege of Boston, Massachusetts furnished 11,500, Connecticut 2,300, New Hampshire 1,200, Rhode Island 1,000. These contingents were arrayed under their local commanders, and under the local flags of their respective commonwealths, though Artemas Ward of Massachusetts had by courtesy exercised the chief command until the arrival of Washington. During the month of July, Congress gave a more continental complexion to the army by sending a reinforcement of 3,000 men from Pennsylvania, Maryland, and Virginia, including the famous Daniel Morgan, with his sturdy band of sharpshooters, each man of whom, it was said, while marching at double-quick, could cleave with his rifle-ball a squirrel at a distance of three hundred yards. The summer of 1775 thus brought together in Cambridge many officers whose names

Washington arrives in Cambridge.

were soon to become household words throughout the length and breadth of the land, and a moment may be fitly spent in introducing them before we proceed with the narrative of events.

Daniel Morgan.

Daniel Morgan, who had just arrived from Virginia with his riflemen, was a native of New Jersey, of Welsh descent. Moving to Virginia at an early age, he had won a great reputation for bravery and readiness of resource in the wild campaigns of the Seven Years' War. He was a man of gigantic stature and strength, and incredible powers of endurance. In his youth, it is said, he had received five hundred lashes by order of a tyrannical British officer, and had come away alive and defiant. On another occasion, in a fierce woodland fight with the Indians, in which nearly all his comrades were slain, Morgan was shot through the neck by a musket-ball. Almost fainting from the wound, which he believed to be fatal, Morgan was resolved, nevertheless, not to leave his scalp in the hands of a dirty Indian; and falling forward, with his arms tightly clasped about the neck of his stalwart horse, though mists were gathering before his eyes, he spurred away through the forest paths, until his foremost Indian pursuer, unable to come up with him, hurled his tomahawk after him with a yell of baffled rage, and gave up the chase. With this unconquerable tenacity, Morgan was a man of gentle and unselfish nature; a genuine diamond, though a rough one; uneducated, but clear and strong in intelligence and faithful in every fibre. At Cambridge began his long comradeship with a very different character, Benedict

Arnold, a young man of romantic and generous impulses, and for personal bravery unsurpassed, but vain and self-seeking, and lacking in moral robustness; in some respects a more polished man than Morgan, but of a nature at once coarser and weaker. We shall see these two men associated in some of the most brilliant achievements of the war; and we shall see them persecuted and insulted by political enemies, until the weaker nature sinks and is ruined, while the stronger endures to the end.

Benedict Arnold.

Along with Morgan and Arnold there might have been seen on Cambridge Common a man who was destined to play no less conspicuous a part in the great campaign which was to end in the first decisive overthrow of the British. For native shrewdness, rough simplicity, and dauntless courage, John Stark was much like Morgan. What the one name was in the great woods of the Virginia frontier, that was the other among the rugged hills of northern New England, — a symbol of patriotism and a guarantee of victory. Great as was Stark's personal following in New Hampshire, he had not, however, the chief command of the troops of that colony. The commander of the New Hampshire contingent was John Sullivan, a wealthy lawyer of Durham, who had sat in the first Continental Congress. Sullivan was a gentleman of wide culture and fair ability as a statesman. As a general, he was brave, intelligent, and faithful, but in no wise brilliant. Closely associated with Sullivan for the next three years we shall find

John Stark.

John Sullivan.

Nathanael Greene, now in command of the Rhode Island contingent. For intellectual calibre the other officers here mentioned are dwarfed at once in comparison with Greene, who comes out at the end of the war with a military reputation scarcely, if at all, inferior to that of Washington. Nor was Greene less notable for the sweetness and purity of his character than for the scope of his intelligence. From lowly beginnings he had come to be, though still a young man, the most admired and respected citizen of Rhode Island. He had begun life as a blacksmith, but, inspired by an intense thirst for knowledge, he had soon become a learned blacksmith, well versed in history, philosophy, and general literature. He had that rare genius which readily assimilates all kinds of knowledge through an inborn correctness of method. Whatever he touched, it was with a master hand, and his weight of sense soon won general recognition. Such a man was not unnaturally an eager book-buyer, and in this way he had some time ago been brought into pleasant relations with the genial and intelligent Henry Knox, who from his bookshop in Boston had come to join the army as a colonel of artillery, and soon became one of Washington's most trusty followers.

Nathanael Greene.

Henry Knox.

Of this group of officers, none have as yet reached very high rank in the Continental army. Sullivan and Greene stand at the end of the list of brigadier-generals; the rest are colonels. The senior major-general, Artemas Ward, and the senior brigadiers, Pomeroy

Older officers.

Heath, Thomas, Wooster, and Spencer, will presently pass into the background, to make way for these younger or more vigorous men. Major-General Israel Putnam, the picturesque wolf-slayer, a brave and sterling patriot, but of slender military capacity, will remain in the foreground for another year, and will then become relegated mainly to garrison duty.

Israel Putnam.

With the exception of Morgan, all the officers here noticed are New England men, as is natural, since the seat of war is in Massachusetts, and an army really continental in complexion is still to be formed. The Southern colonies have as yet contributed only Morgan and the commander-in-chief. New York is represented, in the Continental army, by two of the noblest of American heroes, — Major-General Philip Schuyler and Brigadier-General Richard Montgomery; but these able men are now watching over Ticonderoga and the Indian frontier of New York. But among the group which in 1775 met for consultation on Cambridge Common, or in the noble Tory mansion now hallowed alike by memories of Washington and of Longfellow, there were yet two other generals, closely associated with each other for a time in ephemeral reputation won by false pretences, and afterwards in lasting ignominy. It is with pleasure that one recalls the fact that these men were not Americans, though both possessed estates in Virginia; it is with regret that one is forced to own them as Englishmen. Of Horatio Gates and his career of imbecility and intrigue, we shall by and by see more than enough. At

Horatio Gates and Charles Lee.

this time he was present in Cambridge as adjutant-general of the army. But his friend, Charles Lee, was for the moment a far more conspicuous personage; and this eccentric creature, whose career was for a long time one of the difficult problems in American history, needs something more than a passing word of introduction.

Lee's personal peculiarities.

Although Major-General Charles Lee happened to have acquired an estate in Virginia, he had nothing in common with the illustrious family of Virginian Lees beyond the accidental identity of name. He was born in England, and had risen in the British army to the rank of lieutenant-colonel. He had served in America in the Seven Years' War, and afterward, as a soldier of fortune, he had wandered about Europe, obtaining at one time a place on the staff of the king of Poland. A restless adventurer, he had come over again to America as soon as he saw that a war was brewing here. There is nothing to show that he cared a rush for the Americans, or for the cause in which they were fighting, but he sought the opportunity of making a name for himself. He was received with enthusiasm by the Americans. His loud, pompous manner and enormous self-confidence at first imposed upon everybody. He was tall, lank, and hollow-cheeked, with a discontented expression of face. In dress he was extremely slovenly. He was fond of dogs, and always had three or four at his heels, but toward men and women his demeanour was morose and insulting. He had a sharp, cynical wit, and was always making severe remarks in a harsh, rough voice. But disagreeable as he was, the trustful American imagination endowed him with the qualities of a great soldier. His reputation was part of the unconscious tribute which the provincial mind of our countrymen was long wont to pay to the men and things of Europe; and for some time his worst actions found a lenient interpretation as the mere eccentricities of a wayward genius. He had hoped to be made commander-in-chief of the army, and had already begun to nourish a bitter grudge against Washington, by whom he regarded himself as supplanted. In the following year we shall see him endeavouring to thwart the plans of Washington at the most critical moment of the war, but for the present he showed no signs of insincerity, except perhaps in an undue readiness to parley with the British commanders. As soon as it became clear that a war was beginning, the hope of winning glory by effecting an accommodation with the enemy offered a dangerous temptation to men of weak virtue in eminent positions. In October, 1775, the American camp was thrown into great consternation by the discovery that Dr. Benjamin Church, one of the most conspicuous of the Boston leaders, had engaged in a secret correspondence with the enemy. Dr. Church was thrown into jail, but as the evidence of treasonable intent was not absolutely complete, he was set free in the following spring, and allowed to visit the West Indies for his health. The ship in which he sailed was never heard from again. This kind of temptation, to which Church succumbed at the first outbreak of the war, beset Lee with fatal

Benjamin Church.

effect after the Declaration of Independence, and wrought the ruin of Arnold after the conclusion of the French alliance.

To such a man as Charles Lee, destitute of faith in the loftier human virtues or in the strength of political ideas, it might easily have seemed that more was to be hoped from negotiation than from an attempt to resist Great Britain with such an army as that of which he now came to command the left wing. It was fortunate that the British generals were ignorant of the real state of things. Among the moral effects of the battle of Bunker Hill there was one which proved for the moment to be of inestimable value. It impressed upon General Howe, who now succeeded to the chief command, the feeling that the Americans were more formidable than had been supposed, and that much care and forethought would be required for a successful attack upon them. In a man of his easy-going disposition, such a feeling was enough to prevent decisive action. It served to keep the British force idle in Boston for months, and was thus of great service to the American cause. For in spite of the zeal and valour it had shown, this army of New England minute-men was by no means in a fit condition for carrying on such an arduous enterprise as the siege of Boston. When Washington took command of the army on Cambridge Common, he found that the first and most trying task before him was out of this excellent but very raw material to create an army upon which he could depend. The battle of Bunker Hill had just been lost, under circumstances which were calculated to cheer the Americans and make them hopeful of the future; but it would not do to risk another battle, with an untrained staff and a scant supply of powder. All the work of organizing an army was still to be done, and the circumstances were not such as to make it an easy work. It was not merely that the men, who were much better trained in the discipline of the town-meeting than in that of the camp, needed to be taught the all-important lesson of military subordination: it was at first a serious question how they were to be kept together at all. That the enthusiasm kindled on the day of Lexington should have sufficed to bring together 16,000 men, and to keep them for three months at their posts, was already remarkable; but no army, however patriotic and self-sacrificing, can be supported on enthusiasm alone. The army of which Washington took command was a motley crowd, clad in every variety of rustic attire, armed with trusty muskets and rifles, as their recent exploit had shown, but destitute of almost everything else that belongs to a soldier's outfit. From the Common down to the river, their rude tents were dotted about here and there, some made of sail-cloth stretched over poles, some piled up of stones and turf, some oddly wrought of twisted green boughs; while the more fortunate ones found comparatively luxurious quarters in Massachusetts Hall, or in the little Episcopal church, or in the houses of patriotic citizens. These volunteers had enlisted for various periods, under various contracts with various town or provincial govern-

Difficult work for Washington.

ments. Their terms of service had naturally been conceived to be short, and it was not only not altogether clear how they were going to be paid, but it was not easy to see how they were going to be fed. That this army should have been already subsisted for three months, without any commissariat, was in itself an extraordinary fact. Day by day the heavy carts had rumbled into Cambridge, bringing from the highlands of Berkshire and Worcester, and from the Merrimac and Connecticut valleys, whatever could in any wise be spared of food, or clothing, or medicines, for the patriot army; and the pleasant fields of Cambridge were a busy scene of kindness and sympathy.

Such means as these, however, could not long be efficient. If war was to be successfully conducted, there must be a commissariat, there must be ammunition, and there must be money. And here Washington found himself confronted with the difficulty which never ceased to vex his noble soul and disturb his best laid schemes until the day when he swooped down upon Cornwallis at Yorktown. He had to keep making the army, with which he was too often expected to fight battles ere it was half made; and in this arduous work he could get but little systematic help from any quarter. At present the difficulty was that there was nowhere any organized government competent to support an army. On Washington's arrival, the force surrounding Boston owed allegiance, as we have seen, to four distinct commonwealths, of which two, indeed, — Connecticut and Rhode Island, — pre-

Absence of governmental organization.

serving their ancient charters, with governors elected by themselves, were still in their normal condition. In New Hampshire, on the other hand, the royal governor, Wentworth, whose personal popularity was deservedly great, still kept his place, while Stark and his men had gone to Cambridge in spite of him. In Massachusetts the revolutionary Provincial Congress still survived, but with uncertain powers; even the Continental Congress which adopted the Cambridge army in the name of the United Colonies was simply an advisory body, without the power to raise taxes or to beat up recruits. From this administrative chaos, through which all the colonies, save Connecticut and Rhode Island, were forced to pass in these trying times, Massachusetts was the first to emerge, in July, 1775, by reverting to the provisions of its old charter, and forming a government in which the king's authority was virtually disallowed. A representative assembly was chosen by the people in their town-meetings, according to time-honoured precedent; and this new legislature itself elected an annual council of twenty-eight members, to sit as an upper house. James Bowdoin, as president of the council, became chief executive officer of the commonwealth, and John Adams was made chief-justice. Forty thousand pounds were raised by a direct tax on polls and on real estate, and bills of credit were issued for 1,000 more. The commonwealth adopted a new seal, and a proclamation, issued somewhat later by Chief-Justice Adams, enjoining it upon all people to give loyal

New government of Massachusetts, July, 1775.

obedience to the new government, closed with the significant invocation "God save the people," instead of the customary "God save the king."

In taking this decisive step, Massachusetts was simply the first to act upon the general recommendation of the Continental Congress, that the several colonies should forthwith proceed to frame governments for themselves, based upon the suffrages of the people. From such a recommendation as this to a formal declaration of independence, the distance to be traversed was not great. Samuel Adams urged that in declaring the colonies independent Congress would be simply recognizing a fact which in reality already existed, and that by thus looking facts squarely in the face the inevitable war might be conducted with far greater efficiency. But he was earnestly and ably opposed by Dickinson of Pennsylvania, whose arguments for the present prevailed in the Congress. It was felt that the Congress, as a mere advisory body, had no right to take a step of such supreme importance without first receiving explicit instructions from every one of the colonies. Besides this, the thought of separation was still a painful thought to most of the delegates, and it was deemed well worth while to try the effect of one more candid statement of grievances, to be set forth in a petition to his majesty. For like reasons, the Congress did not venture to take measures to increase its own authority; and when Franklin, still thinking of union as he had been thinking for more than twenty years, now brought forward a new

Congress sends a petition to the king.

scheme, somewhat similar to the Articles of Confederation afterwards adopted, it was set aside as premature. The king was known to be fiercely opposed to any dealings with the colonies as a united body, and so considerate of his feelings were these honest and peace-loving delegates that, after much discussion, they signed their carefully worded petition severally, and not jointly. They signed it as individuals speaking for the people of the American colonies, not as members of an organic body representing the American people. To emphasize still further their conciliatory mood, the delivery of the petition was entrusted to Richard Penn, a descendant of the great Quaker and joint-proprietary in the government of Pennsylvania, an excellent man and an ardent loyalist. At the same time that this was done, an issue of paper money was made, to be severally guaranteed by the thirteen colonies, and half a million dollars were sent to Cambridge to be used for the army.

Military operations, however, came for the time to a stand-still. While Washington's energies were fully occupied in organizing and drilling his troops, in providing them with powder and ball, in raising lines of fortification, in making good the troublesome vacancies due to short terms of enlistment, and above all in presenting unfailingly a bold front to the enemy; while the encampments about Boston were the daily scene of tedious work, without any immediate prospect of brilliant achievement, the Congress and the people were patiently waiting to hear the result of the last petition that was ever to be sent from these colonies to the king of Great Britain.

Penn made all possible haste, and arrived in London on the 14th of August; but when he got there the king would neither see him, nor receive the petition in any way, directly or indirectly. The Congress was an illegal assembly which had no business to send letters to him: if any one of the colonies wanted to make terms for itself separately, he might be willing to listen to it. But this idea of a united America was something unknown either to law or to reason, something that could not be too summarily frowned down. So while Penn waited about London, the king issued a proclamation; setting forth that many of his subjects in the colonies were in open and armed rebellion, and calling upon all loyal subjects of the realm to assist in bringing to condign punishment the authors and abettors of this foul treason. Having launched this thunderbolt, George sent at once to Russia to see if he could hire 20,000 men to aid in giving it effect, for the "loyal subjects of the realm" were slow in coming forward. A war against the Americans was not yet popular in England. Lord Chatham withdrew his eldest son, Lord Pitt, from the army, lest he should be called upon to serve against the men who were defending the common liberties of Englishmen. There was, moreover, in England as well as in America, a distrust of regular armies. Recruiting was difficult, and conscription was something that the people would not endure unless England should actually be threatened with invasion. The king had already been obliged to raise a force of his Hanoverian subjects to garrison

The king issues a proclamation, and tries to hire troops from Russia.

Minorca and Gibraltar, thus setting free the British defenders of these strongholds for service in America. He had no further resource except in hiring troops from abroad. But his attempt in Russia was not successful, for the Empress Catherine, with all her faults, was not disposed to sell the blood of her subjects. She improved the occasion — as sovereigns and others will sometimes do — by asking George, sarcastically, if he thought it quite compatible with his dignity to employ foreign troops against his own subjects; as for Russian soldiers, she had none to spare for such a purpose. Foiled in this quarter, the king applied to the Duke of Brunswick, the Landgrave of Hesse-Cassel, the princes of Waldeck and Anhalt-Zerbst, the Margrave of Anspach-Bayreuth, and the Count of Hesse-Hanau, and succeeded in making a bargain for 20,000 of the finest infantry in Europe, with four good generals, — Riedesel of Brunswick, and Knyphausen, Von Heister, and Donop of Hesse. The hiring of these troops was bitterly condemned by Lord John Cavendish in the House of Commons, and by Lords Camden and Shelburne and the Duke of Richmond in the House of Lords; and Chatham's indignant invectives at a somewhat later date are familiar to every one. It is proper, however, that in such an affair as this we should take care to affix our blame in the right place. The king might well argue that in carrying on a war for what the majority of Parliament regarded as a righteous object, it was no worse for him to hire men than to buy cannon

Catherine refuses.

The king hires German troops.

and ships. The German troops, on their part, might justly complain of Lord Camden for stigmatizing them as "mercenaries," inasmuch as they did not come to America for pay, but because there was no help for it. It was indeed with a heavy heart that these honest men took up their arms to go beyond sea and fight for a cause in which they felt no sort of interest, and great was the mourning over their departure. The persons who really deserved to bear the odium of this transaction were the mercenary princes who thus shamelessly sold their subjects into slavery. It was a striking instance of the demoralization which had been wrought among the petty courts of Germany in the last days of the old empire, and among the German people it excited profound indignation. The popular feeling was well expressed by Schiller, in his "Cabale und Liebe." Frederick the Great, in a letter to Voltaire, declared himself beyond measure disgusted, and by way of publicly expressing his contempt for the transaction he gave orders to his custom-house officers that upon all such of these soldiers as should pass through Prussian territory a toll should be levied, as upon "cattle exported for foreign shambles."

Indignation in Germany.

When the American question was brought up in the autumn session of Parliament, it was treated in the manner with which the Americans had by this time become familiar. A few far-sighted men still urged the reasonableness of the American claims, but there was now a great majority against them. In spite of grave warning voices, both houses decided to support the king; and in this they were upheld by the university of Oxford, which a century ago had burned the works of John Milton as "blasphemous," and which now, with equal felicity, in a formal address to the king, described the Americans as "a people who had forfeited their lives and their fortunes to the justice of the state." At the same time the department of American affairs was taken from the amiable Lord Dartmouth, and given to the truculent Lord George Germain. Those things were done in November, 1775, and in the preceding month they had been heralded by an act of wanton barbarity on the part of a British naval officer, albeit an unwarranted act, which the British government as promptly as possible disowned. On the 16th of October, Captain Mowatt had sailed with four small vessels into the harbour of Portland (then called Falmouth), and with shells and grenades set fire to the little town. St. Paul's Church, all the public buildings, and three fourths of all the dwellings were burned to the ground, and a thousand unoffending men, women, and children were thus turned out-of-doors just as the sharp Maine winter was coming on to starve and freeze them.

Burning of Portland, Oct. 16, 1775.

The news of the burning of Portland reached Philadelphia on the same day (October 31) with the news that George III. was about to send foreign mercenaries to fight against his American subjects; and now the wrath of Congress was thoroughly kindled, and the party which advised further temporizing was thrown into helpless minority.

"Well, brother rebel," said a Southern member to Samuel Ward of Rhode Island, "we have now got a sufficient answer to our petition: I want nothing more, but am ready to declare ourselves independent." Congress now advised New Hampshire, Virginia, and South Carolina to frame for themselves new republican governments, as Massachusetts had already done; it urged South Carolina to seize the British vessels in her waters; it appointed a committee to correspond with foreign powers; and above all, it adopted unreservedly the scheme, already partially carried out, for the expulsion of the British from Canada.

Effects upon Congress.

At once upon the outbreak of hostilities at Lexington, the conquest of Canada had been contemplated by the Northern patriots, who well remembered how, in days gone by, the valley of the St. Lawrence had furnished a base for attacks upon the province of New York, which was then the strategic centre of the American world. It was deemed an act of military prudence to secure this region at the outset. But so long as the least hope of conciliation remained, Congress was unwilling to adopt any measures save such as were purely defensive in character. As we have seen, it was only with reluctance that it had sanctioned the garrisoning of Ticonderoga by the Connecticut troops. But in the course of the summer it was learned that the governor of Canada, Sir Guy Carleton, was about to take steps to recover Ticonderoga; and it was credibly reported that intrigues were going on with the Iro-

The Americans invade Canada, Sept., 1775.

quois tribes, to induce them to harry the New England frontier and the pleasant farms on the Hudson : so that, under these circumstances, the invasion of Canada was now authorized by Congress as a measure of self-defence. An expedition down Lake Champlain, against Montreal, was at once set on foot. As Schuyler, the commander of the northern department, was disabled by ill health, the enterprise was confided to Richard Montgomery, and it could not have been put in better hands. Late in August, Montgomery started from Ticonderoga, and on the 12th of September, with a force of two thousand men, he laid siege to the fortress of St. John's, which commanded the approach to Montreal. Carleton, whose utmost exertions could bring together only some nine hundred men, made heroic but fruitless efforts to stop his progress. After a siege of fifty days, St. John's surrendered on the 3d of November, and on the 12th Montgomery entered Montreal in triumph. The people of Canada had thus far seemed favourably disposed toward the American invaders, and Montgomery issued a proclamation urging them to lose no time in choosing delegates to attend the Continental Congress.

Meanwhile, in September, Washington had detached from the army at Cambridge one thousand New England infantry, with two companies of Pennsylvania riflemen and Morgan's famous Virginia sharpshooters, and ordered them to advance upon Quebec through the forests of Maine and by way of the rivers Kennebec and Chaudière. The expedition was commanded by Colonel Benedict

Arnold, who seems to have been one of the first, if not the first, to suggest it. The enterprise was one to call for all his persistent daring and fertile resource. It was an amphibious journey, as they now rowed their boats with difficulty against the strong, swift current of the Kennebec, and now, carrying boats and oars on their shoulders, forced their way through the tangled undergrowth of the primeval forests. Often they had to wade across perilous bogs, and presently their shoes were cut to pieces by sharp stones, and their clothes torn to shreds by thorns and briers. Their food gave out, and though some small game was shot, their hunger became such that they devoured their dogs. When they reached the head of the Chaudière, after this terrible march of thirty-three days, two hundred of their number had succumbed to starvation, cold, and fatigue, while two hundred more had given out and returned to Massachusetts, carrying with them such of the sick and disabled as they could save. The descent of the Chaudière in their boats afforded some chance for rest, and presently they began to find cattle for food. At last, on the 13th of November, the next day after Montgomery's capture of Montreal, they crossed the broad St. Lawrence, and climbed the Heights of Abraham at the very place where Wolfe had climbed to victory sixteen years ago. There was splendid bravado in Arnold's advancing to the very gates with his little, worn-out army, now reduced to seven hundred men, and summoning the garrison either to come out and fight, or to surren-

Arnold's march through the wilderness of Maine.

der the town. But the garrison very properly would neither surrender nor fight. The town had been warned in time, and Arnold had no alternative but to wait for Montgomery to join him.

Six days afterward, Carleton, disguised as a farmer, and ferried down stream in a little boat, found his way into Quebec; and on the 3d of December, Montgomery made his appearance with a small force, which raised the number of the Americans to twelve hundred men. As Carleton persistently refused to come out of his defences, it was resolved to carry the works by storm, — a chivalrous, nay, one might almost say, a foolhardy decision, had it not been so nearly justified by the event. On the last day of 1775, England came within an ace of losing Quebec. At two o'clock in the morning, in a blinding snowstorm, Montgomery and Arnold began each a furious attack, at opposite sides of the town; and aided by the surprise, each came near carrying his point. Montgomery had almost forced his way in when he fell dead, pierced by three bullets; and this so chilled the enthusiasm of his men that they flagged, until reinforcements drove them back. Arnold, on his side, was severely wounded and carried from the field; but the indomitable Morgan took his place, and his Virginia company stormed the battery opposed to them, and fought their way far into the town. Had the attack on the other side been kept up with equal vigour, as it might have been but for Montgomery's death, Quebec must have fallen. As it was, Morgan's triumphant advance only served to

Assault upon Quebec, Dec. 31, 1775.

isolate him, and presently he and his gallant company were surrounded and captured.

Total failure of the attempt upon Canada.

With the failure of this desperate attack passed away the golden opportunity for taking the citadel of Canada. Arnold remained throughout the winter in the neighbourhood of Quebec, and in the spring the enterprise was taken up by Wooster and Sullivan with fresh forces. But by this time many Hessians had come over, and Carleton, reinforced until his army numbered 13,000, was enabled to recapture Montreal and push back the Americans, until in June, after a hazardous retreat, well conducted by Sullivan, the remnant of their invading army found shelter at Crown Point. Such was the disastrous ending of a campaign which at the outset had promised a brilliant success, and which is deservedly famous for the heroism and skill with which it was conducted. The generalship of Montgomery received the warm approval of no less a critic than Frederick the Great; and the chivalrous bravery of Arnold, both in his march through the wilderness and in the military operations which followed, was such that if a kind fate could then and there have cut the thread of his life, he would have left behind him a sweet and shining memory. As for the attempt to bring Canada into the American union, it was one which had no hope of success save through a strong display of military force. The sixteen years which had elapsed since the victory of Wolfe had not transformed the Canadian of the old *régime* into a free-born Englishman. The question at present for him was only that of a choice of allegiance; and while at first the invaders were favourably received, it soon became apparent that between the Catholic and the Puritan there could be but little real sympathy. The Quebec Act, which legalized Catholic worship in Canada, had done much toward securing England's hold upon this part of her American possessions. And although, in the colourless political condition of this northern province, the capture of Quebec might well have brought it into the American union, where it would gradually have taken on a fresh life, as surely as it has done under British guidance, yet nothing short of such a military occupation could have had any effect in determining its languid preferences.

While Canada was thus freed from the presence of the Continental troops, the British army, on the other hand, was driven from Boston, and New England was cleared of the enemy. During the autumn and winter, Washington had drawn his lines as closely as possible about the town, while engaged in the work of organizing and equipping his army. The hardest task was to collect a sufficient quantity of powder and ball, and to bring together siege-guns. As the season wore on, the country grew impatient, and Washington sometimes had to listen to criticisms like those that were directed against McClellan in Virginia, at the beginning of 1862, or against Grant before Vicksburg, in the spring of 1863. President Hancock, who owned a great deal of property in Boston, urged him to set fire to the town and destroy it, if by so doing he could drive the British to their ships. But Washington had

The siege of Boston.

planned much more wisely. By the 1st of March a great quantity of cannon had been brought in by Henry Knox, some of them dragged on sledges all the way from Ticonderoga, and so at last Washington felt himself prepared to seize upon Dorchester Heights. This position commanded the town and harbour even more effectually than Bunker Hill, and why in all these months General Howe had not occupied it one would find it hard to say. He was bitterly attacked for his remissness by the British newspapers, as was quite natural.

Washington seizes Dorchester Heights, March 4, 1776.

Washington chose for his decisive movement the night of the 4th of March. Eight hundred men led the way, escorting the wagons laden with spades and crowbars, hatchets, hammers, and nails; and after them followed twelve hundred men, with three hundred ox-carts, carrying timbers and bales of hay; while the rear was brought up by the heavy siege-guns. From Somerville, East Cambridge, and Roxbury, a furious cannonade was begun soon after sunset and kept up through the night, completely absorbing the attention of the British, who kept up a lively fire in return. The roar of the cannon drowned every other sound for miles around, while all night long the two thousand Americans, having done their short march in perfect secrecy, were busily digging and building on Dorchester Heights, and dragging their siege-guns into position. Early next morning, Howe saw with astonishment what had been done, and began to realize his perilous situation. The commander of the fleet sent word that unless the Americans could be forthwith dislodged, he could not venture to keep his ships in the harbour. Most of the day was consumed in deciding what should be done, until at last Lord Percy was told to take three thousand men and storm the works. But the slaughter of Bunker Hill had taught its lesson so well that neither Percy nor his men had any stomach for such an enterprise. A violent storm, coming up toward nightfall, persuaded them to delay the attack till next day, and by that time it had become apparent to all that the American works, continually growing, had become impregnable. Percy's orders were accordingly countermanded, and it was decided to abandon the town immediately. It was the sixth anniversary of the day on which Samuel Adams had overawed Hutchinson, and forced him to withdraw his two British regiments from Boston. The work then begun was now consummated by Washington, and from that time forth the deliverance of Massachusetts was complete. Howe caused it at once to be known among the citizens that he was about to evacuate Boston, but he threatened to lay the town in ashes if his troops should be fired on. The selectmen conveyed due information of all this to Washington, who accordingly, secure in the achievement of his purpose, allowed the enemy to depart in peace. By the 17th, the eight thousand troops were all on board their ships, and, taking with them all the Tory citizens, some nine hundred in number, they sailed away for Halifax. Their space did not permit them to carry away their heavy arms, and

The British troops evacuate Boston, March 17, 1776.

their retreat, slow as it was, bore marks of hurry and confusion. In taking possession of the town, Washington captured more than two hundred serviceable cannon, ten times more powder and ball than his army had ever seen before, and an immense quantity of muskets, gun-carriages, and military stores of every sort. Thus was New England set free by a single brilliant stroke, with very slight injury to private property, and with a total loss of not more than twenty lives.

The time was now fairly ripe for the colonies to declare themselves independent of Great Britain. The idea of a separation from the mother-country, which in the autumn had found but few supporters, grew in favour day by day through the winter and spring. The incongruousness of the present situation was well typified by the flag which Washington flung to the breeze on New Year's Day at Cambridge, which was made up of thirteen stripes, to represent the United Colonies, but which retained the cross of St. George in the corner. Thus far, said Benjamin Harrison, they had contrived to "hobble along under a fatal attachment to Great Britain," but the time had come when one must consider the welfare of one's own country first of all. As Samuel Adams said, their petitions had not been heard, and yet had been answered by armies and fleets, and by myrmidons hired from abroad. Nothing had made a greater impression upon the American people than this hiring of German troops. It went farther than any other single cause to ripen their minds for the declaration of

A provisional flag.

Effect of the hiring of "myrmidons."

independence. Many now began to agree with the Massachusetts statesman; and while public opinion was in this malleable condition, there appeared a pamphlet which wrought a prodigious effect upon the people, mainly because it gave terse and vigorous expression to views which every one had already more than half formed for himself.

Thomas Paine had come over to America in December, 1774, and through the favour of Franklin had secured employment as editor of the "Pennsylvania Magazine." He was by nature a dissenter and a revolutionist to the marrow of his bones. Full of the generous though often blind enthusiasm of the eighteenth century for the "rights of man," he was no respecter of the established order, whether in church or state. To him the church and its doctrines meant slavish superstition, and the state meant tyranny. Of crude and undisciplined mind, and quite devoid of scholarship, yet endowed with native acuteness and sagacity, and with no mean power of expressing himself, Paine succeeded in making everybody read what he wrote, and achieved a popular reputation out of all proportion to his real merit. Among devout American families his name is still a name of horror and opprobrium, and uneducated freethinkers still build lecture-halls in honour of his memory, and celebrate the anniversary of his birthday, with speeches full of dismal platitudes. The "Age of Reason," which was the cause of all this singular blessing and banning, contains, amid much crude argument, some sound and sensible criticism, such as is often far exceeded in boldness in the

Thomas Paine.

books and sermons of Unitarian and Episcopalian divines of the present day; but its tone is coarse and dull, and with the improvement of popular education it is fast sinking into complete and deserved oblivion. There are times, however, when such caustic pamphleteers as Thomas Paine have their uses. There are times when they can bring about results which are not so easily achieved by men of finer mould and more subtle intelligence. It was at just such a time, in January, 1776, that Paine published his pamphlet, "Common Sense," on the suggestion of Benjamin Rush, and with the approval of Franklin and of Samuel Adams. The pamphlet is full of scurrilous abuse of the English people, and resorts to such stupid arguments as the denial of the English origin of the Americans. Not one third of the people, *even* of Pennsylvania, are of English descent, argues Paine, as if Pennsylvania had been preëminent among the colonies for its English blood, and not, as in reality, perhaps the least English of all the thirteen save New York. But along with all this stuff there was a sensible and striking statement of the practical state of the case between England and the colonies. The reasons were shrewdly and vividly set forth for looking upon reconciliation as hopeless, and for seizing the present moment to declare to the world what the logic of events was already fast making an accomplished fact. Only thus, it was urged, could the States of America pursue a coherent and well-defined policy, and preserve their dignity in the eyes of the world.

"Common Sense."

It was difficult for the printers, with the clumsy presses of that day, to bring out copies of "Common Sense" fast enough to meet the demand for it. More than a hundred thousand copies were speedily sold, and it carried conviction wherever it went. At the same time, Parliament did its best to reinforce the argument by passing an act to close all American ports, and authorize the confiscation of all American ships and cargoes, as well as of such neutral vessels as might dare to trade with this proscribed people. And, as if this were not quite enough, a clause was added by which British commanders on the high seas were directed to impress the crews of such American ships as they might meet, and to compel them, under penalty of death, to enter the service against their fellow-countrymen. In reply to this edict, Congress, in March, ordered the ports of America to be thrown open to all nations; it issued letters of marque, and it advised all the colonies to disarm such Tories as should refuse to contribute to the common defence. These measures, as Franklin said, were virtually a declaration of war against Great Britain. But before taking the last irrevocable step, the prudent Congress waited for instructions from every one of the colonies.

Fulminations and counter-fulminations.

The first colony to take decisive action in behalf of independence was North Carolina, a commonwealth in which the king had supposed the outlook to be especially favourable for the loyalist party. Recovered in some measure from the turbulence of its earlier days, North Carolina was fast becoming a prosperous community of small planters, and its population had increased so rapidly that it now ranked fourth among the colonies, immediately after Pennsylvania. Since the overthrow of the Pretender at Culloden there had been a great immigration of sturdy Scots from the western Highlands, in which the clans of Macdonald and Macleod were especially represented. The celebrated Flora Macdonald herself, the romantic woman who saved Charles Edward in 1746, had lately come over here and settled at Kingsborough with Allan Macdonald, her husband. These Scottish immigrants also helped to colonize the upland regions of South Carolina and Georgia, and they have powerfully affected the race composition of the Southern people, forming an ancestry of which their descendants may well be proud. Though these Highland clansmen had taken part in the Stuart insurrection, they had become loyal enough to the government of George III., and it was now hoped that with their aid the colonies might be firmly secured, and its neighbours on either side overawed. To this end, in January, Sir Henry Clinton, taking with him 2,000 troops, left Boston and sailed for the Cape Fear river, while a force of seven regiments and ten ships-of-war, under Sir Peter Parker, was ordered from Ireland to coöperate with him. At the same time, Martin, the royal governor, who for safety had retired on board a British ship, carried on negotiations with the Highlanders, until a force of 1,600 men was raised, and, under command of Donald Macdonald, marched down toward the coast to welcome the

The Scots in North Carolina.

Clinton sails for the Carolinas.

arrival of Clinton. But North Carolina had its minute-men as well as Massachusetts, and no sooner was this movement perceived than Colonel Richard Caswell, with 1,000 militia, took up a strong position at the bridge over Moore's Creek, which Macdonald was about to pass on his way to the coast. After a sharp fight of a half hour's duration the Scots were seized with panic, and were utterly routed. Nine hundred prisoners, 2,000 stand of arms, and £15,000 in gold were the trophies of Caswell's victory. The Scottish commander and his kinsman, the husband of Flora Macdonald, were taken and lodged in jail, and thus ended the sway of George III. over North Carolina. The effect of the victory was as contagious as that of Lexington had been in New England. Within ten days 10,000 militia were ready to withstand the enemy, so that Clinton, on his arrival, decided not to land, and stayed cruising about Albemarle Sound, waiting for the fleet under Parker, which did not appear on the scene until May. A provincial congress was forthwith assembled, and instructions were sent to the North Carolina delegates in the Continental Congress, empowering them "to concur with the delegates in the other colonies in declaring independency and forming foreign alliances, reserving to the colony the sole and exclusive right of forming a constitution and laws for it."

The fight at Moore's Creek, Feb. 27, 1776.

North Carolina declares for independence.

At the same time that these things were taking place, the colony of South Carolina was framing for itself a new government, and on the 23d of

March, without directly alluding to independence, it empowered its delegates to concur in any measure which might be deemed essential to the welfare of America. In Georgia the provincial congress, in choosing a new set of delegates to Philadelphia, authorized them to "join in any measure which they might think calculated for the common good."

Action of South Carolina and Georgia.

In Virginia the party in favour of independence had been in the minority, until, in November, 1775, the royal governor, Lord Dunmore, had issued a proclamation, offering freedom to all such negroes and indented white servants as might enlist for the purpose of "reducing the colony to a proper sense of its duty." This measure Lord Dunmore hoped would "oblige the rebels to disperse, in order to take care of their families and property." But the object was not attained. The relations between master and slave in Virginia were so pleasant that the offer of freedom fell upon dull, uninterested ears. With light work and generous fare, the condition of the Virginia negro was a happy one. The time had not yet come when he was liable to be torn from wife and children, to die of hardship in the cotton-fields and rice-swamps of the far South. He was proud of his connection with his master's estate and family, and had nothing to gain by rebellion. As for the indented white servants, the governor's proposal to them was of about as much consequence as a proclamation of Napoleon's would have been if, in 1805, he had offered to set free the prisoners in Newgate on

Virginia: Lord Dunmore's proclamation.

condition of their helping him to invade England. But, impotent as this measure of Lord Dunmore's was, it served to enrage the people of Virginia, setting their minds irretrievably against the king and his cause. During the month of November, hearing that a party of "rebels" were on their way from North Carolina to take possession of Norfolk, Lord Dunmore built a rude fort at the Great Bridge over Elizabeth river, which commanded the southern approach to the town. At that time, Norfolk, with about 9,000 inhabitants, was the principal town in Virginia, and the commercial centre of the colony. The loyalist party, represented chiefly by Scottish merchants, was so strong there and so violent that many of the native Virginia families, finding it uncomfortable to stay in their homes, had gone away into the country. The patriots, roused to anger by Dunmore's proclamation, now resolved to capture Norfolk, and a party of sharpshooters, with whom the illustrious John Marshall served as lieutenant, occupied the bank of Elizabeth river, opposite Dunmore's fort. On the 9th of December, after a sharp fight of fifteen minutes, in which Dunmore's regulars lost sixty-one men, while not a single Virginian was slain, the fort was hastily abandoned, and the road to Norfolk was laid open for the patriots. A few days later the Virginians took possession of their town, while Dunmore sought refuge in the Liverpool, ship-of-the-line, which had just sailed into the harbour. On New Year's Day the governor vindictively set fire to the town, which he had been

Skirmish at the Great Bridge; and burning of Norfolk.

unable to hold against its rightful owners. The conflagration, kindled by shells from the harbour, raged for three days and nights, until the whole town was laid in ashes, and the people were driven to seek such sorry shelter as might save them from the frosts of midwinter.)

This event went far toward determining the attitude of Virginia. In November the colony had not felt ready to comply with the recommendation of Congress, and frame for herself a new government. The people were not yet ready to sever the links which bound them to Great Britain. But the bombardment of their principal town was an argument of which every one could appreciate the force and the meaning. During the winter and spring the revolutionary feeling waxed in strength daily. On the 6th of May, 1776, a convention was chosen to consider the question of independence. Mason, Henry, Pendleton, and the illustrious Madison took part in the discussion, and on the 14th it was unanimously voted to instruct the Virginia delegates in Congress "to propose to that respectable body to declare the United Colonies free and independent States," and to "give the assent of the colony to measures to form foreign alliances and a confederation, provided the power of forming government for the internal regulations of each colony be left to the colonial legislatures." At the same time, it was voted that the people of Virginia should establish a new government for their commonwealth. In the evening, when these decisions had been made known to the people of

Virginia declares for independence.

Williamsburgh, their exultation knew no bounds. While the air was musical with the ringing of church-bells, guns were fired, and the British flag was hauled down at the State House, and the thirteen stripes hoisted in its place.

This decisive movement of the largest of the colonies was hailed throughout the country with eager delight; and from other colonies which had not yet committed themselves responses came quickly. Rhode Island, which had never parted with its original charter, did not need to form a new government, but it had already, on the 4th of May, omitted the king's name from its public documents and sheriff's writs, and had agreed to concur with any measures which Congress might see fit to adopt regarding the relations between England and America. In the course of the month of May town-meetings were held throughout Massachusetts and it was everywhere unanimously voted to uphold Congress in the declaration of independence which it was now expected to make.

Action of Rhode Island and Massachusetts.

On the 15th of May, Congress adopted a resolution recommending to all the colonies to form for themselves independent governments, and in a preamble, written by John Adams, it was declared that the American people could no longer conscientiously take oath to support any government deriving its authority from the Crown; all such governments must now be suppressed, since the king had withdrawn his protection from the inhabitants of the United Colonies. Like the famous preamble to Towns-

Resolution of May 15.

hend's bill of 1767, this Adams preamble contained within itself the gist of the whole matter. To adopt it was virtually to cross the Rubicon, and it gave rise to a hot debate. James Duane of New York admitted that if the facts stated in the preamble should turn out to be true, there would not be a single voice against independence; but he could not yet believe that the American petitions were not destined to receive a favourable answer. "Why," therefore, "all this haste? Why this urging? Why this driving?" James Wilson of Pennsylvania, one of the ablest of all the delegates in that revolutionary body, urged that Congress had not yet received sufficient authority from the people to justify it in taking so bold a step. The resolution was adopted, however, preamble and all; and now the affair came quickly to maturity. "The Gordian knot is cut at last!" exclaimed John Adams. In town-meeting the people of Boston thus instructed their delegates: "The whole United Colonies are upon the verge of a glorious revolution. We have seen the petitions to the king rejected with disdain. For the prayer of peace he has tendered the sword; for liberty, chains; for safety, death. Loyalty to him is now treason to our country. We think it absolutely impracticable for these colonies to be ever again subject to or dependent upon Great Britain, without endangering the very existence of the state. Placing, however, unbounded confidence in the supreme council of the Congress, we are determined to wait, most patiently wait, till their wisdom shall

Instructions from Boston.

dictate the necessity of making a declaration of independence. In case the Congress should think it necessary for the safety of the United Colonies to declare them independent of Great Britain, the inhabitants, with their lives and the remnant of their fortunes, will most cheerfully support them in the measure."

This dignified and temperate expression of public opinion was published in a Philadelphia evening paper, on the 8th of June. On the preceding day, in accordance with the instructions which had come from Virginia, the following motion had been submitted to Congress by Richard Henry Lee: —

Lee's motion in Congress.

"That these United Colonies are, and of right ought to be, free and independent States; that they are absolved from all allegiance to the British Crown; and that all political connection between them and the state of Great Britain is, and ought to be, totally dissolved.

"That it is expedient forthwith to take the most effectual measures for forming foreign alliances.

"That a plan of confederation be prepared and transmitted to the respective colonies, for their consideration and approbation."

In these trying times the two greatest colonies, Virginia and Massachusetts, had been wont to go hand in hand; and the motion of Richard Henry Lee was now promptly seconded by John Adams. It was resisted by Dickinson and Wilson of Pennsylvania, and by Robert Livingston of New York, on the ground that public opinion in the middle colonies was not yet ripe for supporting such a measure; at the same time these cautious members freely acknowledged that the lingering hope of an amicable settlement with Great Britain had come to be quite chimerical. The prospect of securing European alliances was freely discussed. The supporters of the motion urged that a declaration of independence would be nothing more than the acknowledgment of a fact which existed already; and until this fact should be formally acknowledged, it was not to be supposed that diplomatic courtesy would allow such powers as France or Spain to treat with the Americans. On the other hand, the opponents of the motion argued that France and Spain were not likely to look with favour upon the rise of a great Protestant power in the western hemisphere, and that nothing would be easier than for these nations to make a bargain with England, whereby Canada might be restored to France and Florida to Spain, in return for military aid in putting down the rebellious colonies. The result of the whole discussion was decidedly in favour of a declaration of independence; but to avoid all appearance of undue haste, it was decided, on the motion of Edward Rutledge of South Carolina, to postpone the question for three weeks, and invite the judgment of those colonies which had not yet declared themselves.

Debate on Lee's motion.

Under these circumstances, the several colonies acted with a promptness that outstripped the expectations of Congress. Connecticut had no need of a new government, for, like Rhode Island, she had always kept the

Connecticut and New Hampshire.

charter obtained from Lord Clarendon in 1662, she had always chosen her own governor, and had always been virtually independent of Great Britain. Nothing now was necessary but to omit the king's name from legal documents and commercial papers, and to instruct her delegates in Congress to support Lee's motion; and these things were done by the Connecticut legislature on the 14th of June. The very next day, New Hampshire, which had formed a new government as long ago as January, joined Connecticut in declaring for independence.

In New Jersey there was a sharp dispute. The royal governor, William Franklin, had a strong party in the colony; and the assembly had lately instructed its delegates to vote against independence, and had resolved to send a separate petition to the king. Against so rash and dangerous a step, Dickinson, Jay, and Wythe were sent by Congress to remonstrate; and as the result of their intercession, the assembly, which yielded, was summarily prorogued by the governor. A provincial congress was at once chosen in its stead. On the 16th of June, the governor was arrested and sent to Connecticut for safe-keeping; on the 21st, it was voted to frame a new government; and on the 22d, a new set of delegates were elected to Congress, with instructions to support the declaration of independence.

New Jersey.

In Pennsylvania there was hot discussion, for the whole strength of the proprietary government was thrown into the scale against independence. Among the Quakers, too, there

Pennsylvania and Delaware.

was a strong disposition to avoid an armed conflict, on any terms. A little while before, they had held a convention, in which it was resolved that "the setting up and putting down kings and governments is God's peculiar prerogative, for causes best known to himself, and that it is not our business to have any hand or contrivance therein; nor to be busybodies above our station, much less to plot and contrive the ruin or overturn of any of them, but to pray for the king and safety of our nation and good of all men; that we may lead a peaceable and quiet life in all goodness and honesty, under the government which God is pleased to set over us. May we, therefore, firmly unite in the abhorrence of all such writings and measures as evidence a desire and design to break off a happy connection we have hitherto enjoyed with the kingdom of Great Britain, and our just and necessary subordination to the king and those who are lawfully placed in authority under him." This view of the case soon met with a pithy rejoinder from Samuel Adams, who, with a quaint use of historical examples, proved that, as the rise of kings and empires is part of God's special prerogative, the time had now come, in the course of divine providence, for the setting up of an independent empire in the western hemisphere. Six months ago, the provincial assembly had instructed its delegates to oppose independence; but on the 20th of May a great meeting was held at the State House, at which more than seven thousand people were present, and it was unanimously resolved that this act of the assembly "had the dangerous tendency to withdraw this province from that happy union with the other colonies which we consider both our glory and our protection." The effect of this resolution was so great that on the 18th of June a convention was held to decide on the question of independence; and after six days of discussion, it was voted that a separation from Great Britain was desirable, provided only that, under the new federal government, each state should be left to regulate its own internal affairs. On the 14th of June, a similar action had been taken by Delaware.

In Maryland there was little reason why the people should wish for a change of government, save through their honourable sympathy with the general interests of the United Colonies. Not only was the proprietary government deeply rooted in the affections of the people, but Robert Eden, the governor holding office at this particular time, was greatly loved and respected. Maryland had not been insulted by the presence of troops. She had not seen her citizens shot down in cold blood like Massachusetts, or her chief city laid in ashes like Virginia; nor had she been threatened with invasion and forced to fight in her own defence like North Carolina. Her direct grievances were few and light, and even so late as the 21st of May, she had protested against any action which might lead to the separation of the colonies from England. But when, in June, her great leaders, Samuel Chase and Charles Carroll of Carrollton, determined to "take the sense of the people," a series of county meetings were

Maryland.

held, and it was unanimously voted that "the true interests and substantial happiness of the United Colonies in general, and this in particular, are inseparably interwoven and linked together." As soon as the colony had taken its stand upon this broad and generous principle, the governor embarked on a British man-of-war before Annapolis, bearing with him the kindly regrets and adieus of the people, and on the 28th of June the delegates in Congress were duly authorized to concur in a declaration of independence.

Peaceful Maryland was thus the twelfth colony which formally committed itself to the cause of independence, as turbulent North Carolina, under the stimulus of civil war and threatened invasion, had been the first. Accordingly on the 1st of July, the day when the motion of Richard Henry Lee was to be taken up in Congress, unanimous instructions in favour of independence had been received from every one of the colonies, except New York. In approaching this momentous question New York was beset by peculiar difficulties. Not only was the Tory party unusually strong there, for reasons already stated, but the risks involved in a revolutionary policy were greater than anywhere else. From its commanding military position, it was clear that the British would direct their main efforts toward the conquest of this central colony; and while on the one hand the broad, deep waters about Manhattan Island afforded an easy entrance for their resistless fleet, on the other hand the failure of the Canadian expedition had laid the whole country open to inva-

The situation in New York.

sion from the north, and the bloodthirsty warriors of the Long House were not likely to let slip so excellent an opportunity for gathering scalps from the exposed settlements on the frontier. Not only was it probable, for these reasons, that New York would suffer more than any other colony from the worst horrors of war, but as a commercial state with only a single seaport, the very sources of her life would be threatened should the British once gain a foothold upon Manhattan Island. The fleet of Lord Howe was daily expected in the harbour, and it was known that the army which had been ousted from Boston, now largely reinforced, was on its way from Halifax to undertake the capture of the city of New York. To guard against this expected danger, Washington had some weeks since moved his army thither from Boston; but his whole effective force did not exceed eight thousand men, and with these he was obliged to garrison points so far apart as King's Bridge, Paulus Hook, Governor's Island, and Brooklyn Heights. The position was far less secure than it had been about Boston, for British ships could here come up the Hudson and East rivers, and interpose between these isolated detachments. As for Staten Island, Washington had not troops enough to occupy it at all, so that when General Howe arrived, on the 28th of June, he was allowed to land there without opposition. It was a bitter thing for Washington to be obliged to permit this, but there was no help for it. Not only in numbers, but in equipment, Washington's force was utterly inadequate to the important task assigned it, and Congress had done nothing to increase its efficiency beyond ordering a levy of twenty-five thousand militia from New England and the middle colonies, to serve for six months only.

Under these circumstances, the military outlook, in case the war were to go on, was certainly not encouraging, and the people of New York might well be excused for some tardiness in committing themselves irrevocably on the question of independence, especially as it was generally understood that Lord Howe was coming armed with plenary authority to negotiate with the American people. To all the other dangers of the situation there was added that of treachery in the camp. Governor Tryon, like so many of the royal governors that year, had taken refuge on shipboard, whence he schemed and plotted with his friends on shore. A plan was devised for blowing up the magazines and seizing Washington, who was either to be murdered or carried on board ship to be tried for treason, according as the occasion might suggest. The conspiracy was discovered in good time; the mayor of New York, convicted of correspondence with Tryon, was thrown into jail, and one of Washington's own guard, who had been bribed to aid the nefarious scheme, was summarily hanged in a field near the Bowery. Such a discovery as this served to throw discredit upon the Tory party. The patriots took a bolder stand than ever, but when the 1st of July came it found the discussion still going on, and the New York delegates in Congress were still without instructions.

The Tryon plot, June, 1776.

On the 1st of July Congress resolved itself into a committee of the whole, to "take into consideration the resolution respecting independency." As Richard Henry Lee was absent, John Adams, who had seconded the motion, was called upon to defend it, which he did in a powerful speech. He was ably opposed by John Dickinson, who urged that the country ought not to be rashly committed to a position, to recede from which would be infamous, while to persist in it might entail certain ruin. A declaration of independence would not strengthen the resources of the country by a single regiment or a single cask of powder, while it would shut the door upon all hope of accommodation with Great Britain. And as to the prospect of an alliance with France and Spain, would it not be well to obtain some definite assurances from these powers before proceeding to extremities? Besides all this, argued Dickinson, the terms of confederation among the colonies were still unsettled, and any declaration of independence, to have due weight with the world, ought to be preceded by the establishment of a federal government. The boundaries of the several colonies ought first to be fixed, and their respective rights mutually guaranteed; and the public lands ought also to be solemnly appropriated for the common benefit. Then, the orator concluded, "when things shall have been thus deliberately rendered firm at home and favourable abroad, — then let America, *attollens humeris famam et fata nepotum*, bearing up her glory and the destiny of her descendants, advance with majestic steps, and assume her station among the sovereigns of the world."

Final debate on Lee's motion.

That there was great weight in some of these considerations was shown only too plainly by subsequent events. But the argument as a whole was open to the fatal objection that if the American people were to wait for all these great questions to be settled before taking a decisive step, they would never be able to take a decisive step at all. The wise statesman regards half a loaf as better than no bread. Independent action on the part of all the colonies except New York had now become an accomplished fact. All were really in rebellion, and their cause could not fail to gain in dignity and strength by announcing itself to the world in its true character. Such was now the general feeling of the committee. When the question was put to vote, the New York delegates were excused, as they had no sufficient instructions. Of the three delegates from Delaware, one was absent, one voted yea, and one nay, so that the vote of the colony was lost. Pennsylvania declared in the negative by four votes against three. South Carolina also declared in the negative, but with the intimation from Edward Rutledge that it might not unlikely reverse its vote, in deference to the majority. The other nine colonies all voted in the affirmative, and the resolution was reported as agreed to by a two thirds vote. On the next day, when the vote was formally taken in regular session of Congress, the Delaware members were all present, and the affirmative vote of that colony was secured; Dickinson and Morris stayed away, thus reversing the vote of Pennsylvania; and the South Carolina members changed for the sake of unanimity.

Vote on Lee's motion.

Thus was the Declaration of Independence at last resolved upon, by the unanimous vote of twelve colonies, on the 2d of July, 1776; and this work having been done, Congress at once went into committee of the whole, to consider the form of declaration which should be adopted. That no time might be lost in disposing of this important matter, a committee had already been selected three weeks before, at the time of Lee's motion, to draw up a paper which might be worthy of this great and solemn occasion. Thomas Jefferson, John Adams, Benjamin Franklin, Roger Sherman, and Robert Livingston were the members of the committee, and Jefferson, as representing the colony which had introduced the resolution of independence, was chosen to be the author of the Declaration. Jefferson, then but thirty-three years of age, was one of the youngest delegates in Congress; but of all the men of that time, there was, perhaps, none of wider culture or keener political instincts. Inheriting a comfortable fortune, he had chosen the law as his profession, but he had always been passionately fond of study for its own sake, and to a very wide reading in history and in ancient and modern literature he added no mean proficiency in mathematics and in physical science. He was skilled in horsemanship and other manly exercises, and in the management of rural affairs; while at the same time he was very sensitively and delicately organized, playing the violin like a master, and giving other evidences of rare musical talent. His temper was exceedingly placid, and his disposition was sweet and

Thomas Jefferson.

sympathetic. He was deeply interested in all the generous theories of the eighteenth century concerning the rights of man and the perfectibility of human nature; and, like most of the contemporary philosophers whom he admired, he was a sturdy foe to intolerance and priestcraft. He was in his way a much more profound thinker than Hamilton, though he had not such a constructive genius as the latter; as a political leader he was superior to any other man of his age; and his warm sympathies, his almost feminine tact, his mastery of the dominant political ideas of the time, and, above all, his unbounded faith in the common sense of the people and in their essential rectitude of purpose served to give him one of the greatest and most commanding positions ever held by any personage in American history.

Independence declared, July 4, 1776.

On the evening of the 4th of July, 1776, the Declaration of Independence was unanimously adopted by twelve colonies, the delegation from New York still remaining unable to act. But the acquiescence of that colony was so generally counted upon that there was no drawback to the exultation of the people. All over the country the Declaration was received with bonfires, with the ringing of bells and the firing of guns, and with torchlight processions. Now that the great question was settled there was a general feeling of relief. "The people," said Samuel Adams, "seem to recognize this resolution as though it were a decree promulgated from heaven." On the 9th of July it was formally adopted by New York, and the soldiers there celebrated the occasion by throwing down the leaden statue of George III. on the Bowling Green, and casting it into bullets.

Thus, after eleven years of irritation, and after such temperate discussion as befitted a free people, the Americans had at last entered upon the only course that could preserve their self-respect, and guarantee them in the great part which they had to play in the drama of civilization. For the dignity, patience, and moderation with which they had borne themselves throughout these trying times, history had as yet scarcely afforded a parallel.

The Declaration was a deliberate expression of the sober thought of the American people.

So extreme had been their forbearance, so great their unwillingness to appeal to brute force while there yet remained the slightest hope of a peaceful solution, that some British historians have gone quite astray in interpreting their conduct. Because statesmen like Dickinson and communities like Maryland were slow in believing that the right moment for a declaration of independence had come, the preposterous theory has been suggested that the American Revolution was the work of an unscrupulous and desperate minority, which, through intrigue mingled with violence, succeeded in forcing the reluctant majority to sanction its measures. Such a misconception has its root in an utter failure to comprehend the peculiar character of American political life, like the kindred misconception which ascribes the rebellion of the colonies to a sordid unwillingness to bear their due share of the expenses of the British Empire. It is like the misunderstanding which saw an angry mob in every town-meeting of the people of Boston, and characterized as a "riot" every deliberate expression of public opinion. No one who is familiar with the essential features of American political life can for a moment suppose that the Declaration of Independence was brought about by any less weighty force than the settled conviction of the people that the priceless treasure of self-government could be preserved by no other means. It was but slowly that this unwelcome conviction grew upon the people; and owing to local differences of circumstances it grew more slowly in some places than in others. Prescient leaders, too, like the Adamses and Franklin and Lee, made up their minds sooner than other people. Even those conservatives who resisted to the last, even such men as John Dickinson and Robert Morris, were fully agreed with their opponents as to the principle at issue between Great Britain and America, and nothing would have satisfied them short of the total abandonment by Great Britain of her pretensions to impose taxes and revoke charters. Upon this fundamental point there was very little difference of opinion in America. As to the related question of independence, the decision, when once reached, was everywhere alike the reasonable result of free and open discussion; and the best possible illustration of this is the fact that not even in the darkest days of the war already begun did any state deliberately propose to reconsider its action in the matter. The hand once put to the plough, there was no turning back. As Judge Drayton of South Carolina said from the bench, " A decree is now gone forth not to be recalled, and thus has suddenly risen in the world a new empire, styled the United States of America."

CHAPTER V.

FIRST BLOW AT THE CENTRE.

THROUGHOUT a considerable portion of the country the news of the Declaration of Independence was accompanied by the news of a brilliant success at the South. After the defeat of Macdonald at Moore's Creek, and the sudden arming of North Carolina, Clinton did not venture to land, but cruised about in the neighbourhood, awaiting the arrival of Sir Peter Parker's squadron from Ireland. Harassed by violent and contrary winds, Parker was three months in making the voyage, and it was not until May that he arrived, bringing with him Lord Cornwallis. As North Carolina had given such unmistakable evidence of its real temper, it was decided not to land upon that coast for the present, but to go South and capture Charleston and Savannah. Lord William Campbell, refugee governor of South Carolina, urged that there was a great loyalist party in that colony, which would declare itself as soon as the chief city should be in the hands of the king's troops. That there would be any serious difficulty in taking Charleston occurred to no one. But Colonel Moultrie had thrown up on Sullivan's Island, commanding the harbour, a fortress of palmetto logs strength-

Lord Cornwallis arrives upon the scene.

ened by heavy banks of sand, and now held it with a force of twelve hundred men, while five thousand militia were gathered about the town, under command of General Charles Lee, who had been sent down to meet the emergency, but did little more than to meddle and hinder. In his character of trained European officer, Lee laughed to scorn Moultrie's palmetto stronghold, and would have ordered him to abandon it, but that he was positively overruled by Rutledge, president of the provincial congress, who knew Moultrie and relied upon his sound judgment. The British commanders, Clinton and Parker, wasted three weeks in discussing various plans of attack, while the Americans, with spade and hatchet, were rapidly barring every approach to Charleston, and fresh regiments came pouring in to man the new-built intrenchments. At last Clinton landed three thousand men on a naked sand-bank, divided from Sullivan's Island by a short space of shallow sea, which he thought could be forded at low tide. At the proper time Sir Peter Parker was to open a lively fire from the fleet, which it was expected would knock down the fort in a few minutes, while Clinton, fording the shoals, would drive out the Americans at the point of the bayonet. The shoals, however, turned out to be seven feet deep at low water, and the task of the infantry was reduced to a desperate conflict with the swarms of mosquitoes, which nearly drove them frantic. The battle thus became a mere artillery duel between the fort and the fleet. The British fire was rapid and furious, but inef-

Battle of Fort Moultrie, June 28, 1776.

fective. Most of the shot passed harmlessly over the low fortress, and those which struck did no harm to its elastic structure. The American fire was very slow, and few shots were wasted. The cable of Parker's flagship was cut by a well-aimed ball, and the ship, swinging around, received a raking fire which swept her deck with terrible slaughter. After the fight had lasted ten hours, the British retreated out of range. The palmetto fort had suffered no serious injury, and only one gun had been silenced. The American loss in killed and wounded was thirty-seven. On the other hand, Sir Peter's flagship had lost her main-mast and mizzen-mast, and had some twenty shots in her hull, so that she was little better than a wreck. The British loss in killed and wounded was two hundred and five. Of their ten sail, only one frigate remained seaworthy at the close of the action. After waiting three weeks to refit, the whole expedition sailed away for New York to coöperate with the Howes. Charleston was saved, and for more than two years the southern states were freed from the invader. In commemoration of this brilliant victory, and of the novel stronghold which had so roused the mirth of the European soldier of fortune, the outpost on Sullivan's Island has ever since been known by the name of Fort Moultrie.

It was with such tidings of good omen that the Declaration of Independence was sent forth to the world. But it was the last news of victory that for the next six months was to cheer the anxious statesmen assembled at Philadelphia. During the rest of the summer and the autumn, disaster followed upon disaster, until it might well seem as if fickle fortune had ceased to smile upon the cause of liberty. The issue of the contest was now centred in New York. By conquering and holding the line of the Hudson river, the British hoped to cut the United Colonies in two, after which it was thought that Virginia and New England, isolated from each other, might be induced to consider the error of their ways and repent. Accordingly, General Howe was to capture the city of New York, while General Carleton was to descend from Canada, recapture Ticonderoga, and take possession of the upper waters of the Hudson, together with the Mohawk valley. Great hopes were built upon the coöperation of the loyalists, of whom there was a greater number in New York than in any other state, except perhaps South Carolina. It was partly for this reason, as we shall hereafter see, that these two states suffered more actual misery from the war than all the others put together. The horrors of civil war were to be added to the attack of the invader. Throughout the Mohawk valley the influence of Sir John Johnson, the Tory son of the famous baronet of the Seven Years' War, was thought to be supreme; and it turned out to be very powerful both with the white population and with the Indians. At the other end of the line, in New York city, the Tory element was strong, for reasons already set forth. On Long Island, the people of Kings and Queens counties, of Dutch

British plan for conquering the Hudson, and cutting the United Colonies in twain.

descent, were Tories almost to a man, while the English population of Suffolk was solidly in favour of independence. And this instance of Long Island was typical. From one end of the United States to the other, as might have been expected, the Tory sentiment was strongest with the non-English element in the population.

Before beginning his attack on New York, General Howe had to await the arrival of his brother; for the ministry had resolved to try the effect of what seemed to them a "conciliatory policy." On the 12th of July Lord Howe arrived at Staten Island, bringing with him the "olive-branch" which Lord North had promised to send along with the sword. This curious specimen of political botany turned out to consist of a gracious declaration that all persons who should desist from rebellion and lend their "aid in restoring tranquillity" would receive full and free pardon from their sovereign lord the king. As it would not do to recognize the existence of Congress, Lord Howe inclosed this declaration in a letter addressed to "George Washington, Esq.," and sent it up the harbour with a flag of truce. But as George Washington, in his capacity of Virginian landholder and American citizen, had no authority for dealing with a royal commissioner, he refused to receive the letter. Colonel Reed informed Lord Howe's messenger that there was no person in the army with that address. The British officer reluctantly rowed away, but suddenly, putting his barge about, he came back and inquired by what title Washington should be

Lord Howe's futile attempt to negotiate with Washington unofficially.

properly addressed. Colonel Reed replied, "You are aware, sir, of the rank of General Washington in our army?" "Yes, sir, we are," answered the officer; "I am sure my Lord Howe will lament exceedingly this affair, as the letter is of a civil, and not of a military nature. He greatly laments that he was not here a little sooner." This remark was understood by Colonel Reed to refer to the Declaration of Independence, which was then but eight days old. A week later Lord Howe sent Colonel Patterson, the British adjutant-general, with a document now addressed to "George Washington, Esq., etc., etc." Colonel Patterson begged for a personal interview, which was granted. He was introduced to Washington, whom he describes as a gentleman of magnificent presence and very handsomely dressed. Somewhat overawed, and beginning his remarks with "May it please your Excellency," Patterson explained that the etceteras on the letter meant everything. "Indeed," said Washington, with a pleasant smile, "they might mean anything." He declined to take the letter, but listened to Patterson's explanations, and then replied that he was not authorized to deal with the matter, and could not give his lordship any encouragement, as he seemed empowered only to grant pardons, whereas those who had committed no fault needed no pardons. As Patterson got up to go, he asked if his Excellency had no message to send to Lord Howe. "Nothing," answered Washington, "but my particular compliments." Thus foiled in his attempt to negotiate with the American commander, Lord

Howe next inclosed his declaration in a circular letter addressed to the royal governors of the middle and southern colonies; but as most of these dignitaries were either in jail or on board the British fleet, not much was to be expected from such a mode of publication. The precious document was captured and sent to Congress, which derisively published it for the amusement and instruction of the people. It was everywhere greeted with jeers. "No doubt we all need pardon from Heaven," said Governor Trumbull of Connecticut, "for our manifold sins and transgressions; but the American who needs the pardon of his Britannic Majesty is yet to be found." The only serious effect produced was the weakening of the loyalist party. Many who had thus far been held back by the hope that Lord Howe's intercession might settle all the difficulties now came forward as warm supporters of independence as soon as it became apparent that the king had really nothing to offer.

The olive-branch having proved ineffectual, nothing was left but to unsheathe the sword, and an interesting campaign now began, of which the primary object was to capture the city of New York and compel Washington's army to surrender. The British army was heavily reinforced by the return of Clinton's expedition and the arrival of 11,000 fresh troops from England and Germany. General Howe had now more than 25,000 men at his disposal, fully equipped and disciplined; while to oppose him Washington had but 18,000, many of

The military problem at New York.

them raw levies which had just come in. If the American army had consisted of such veterans as Washington afterwards led at Monmouth, the disparity of numbers would still have told powerfully in favour of the British. As it was, in view of the crudeness of his material, Washington could hardly hope to do more with his army than to make it play the part of a detaining force. To keep the field in the face of overwhelming odds is one of the most arduous of military problems, and often calls for a higher order of intelligence than that which is displayed in the mere winning of battles. Upon this problem Washington was now to be employed for six months without respite, and it was not long before he gave evidence of military genius such as has seldom been surpassed in the history of modern warfare. At the outset the city of New York furnished the kernel of the problem. Without control of the water it would be well-nigh impossible to hold the city. Still there was a chance, and it was the part of a good general to take this chance, and cut out as much work as possible for the enemy. The shore of Manhattan Island was girded with small forts and redoubts, which Lee had erected in the spring before his departure for South Carolina. The lower end of the island, along the line of Wall Street, was then but little more than half its present width, as several lines of street have since been added upon both sides. From Cortlandt Street across to Paulus Hook, the width of the Hudson river was not less than two miles, while the East river near Fulton Ferry was nearly a mile in width. The city reached only from the Battery as far as Chatham Street, whence the Bowery Lane ran northwestwardly to Bloomingdale through a country smiling with orchards and gardens. Many of the streets were now barricaded, and a strong line of redoubts ran across from river to river below the side of Canal Street. At the upper end of the island, and on the Jersey shore, were other fortresses, with which we shall shortly have to deal, and out in the harbour, as a sort of watch-tower from which to inspect the enemy's fleet, a redoubt had been raised on Governor's Island, and was commanded by Colonel Prescott, with a party of the men of Bunker Hill.

In order to garrison such various positions, it was necessary for Washington to scatter his 18,000 men; and this added much to the difficulty of his task, for Howe could at any moment strike at almost any one of these points with his whole force. From the nature of the case the immense advantage of the initiative belonged entirely to Howe. But in one quarter, the most important of all, Washington had effected as much concentration of his troops as was possible. The position on Brooklyn Heights was dangerously exposed, but it was absolutely necessary for the Americans to occupy it if they were to keep their hold upon New York. This eminence commanded New York exactly as Bunker Hill and Dorchester Heights commanded Boston. Greene had, accordingly, spent the summer in fortifying it, and there 9,000 men — one half of the army — were now concentrated under command of Putnam.

Importance of Brooklyn Heights.

Upon this exposed position General Howe determined to throw nearly the whole of his force. He felt confident that the capture or destruction of half the American army would so discourage the rebels as to make them lend a readier ear to the overtures of that excellent peacemaker, his brother. Accordingly, on the 22d of August, General Howe landed 20,000 men at Gravesend Bay. From this point the American position was approachable by four roads, two of which crossed a range of densely wooded hills, and continued through the villages of Bedford and Flatbush. To the left of these the Gowanus road followed the shore about the western base of the hills, while on the right the Jamaica road curved inland and turned their eastern base.

The elaborate caution with which the British commander now proceeded stands out in striking contrast with the temerity of his advance upon Bunker Hill in the preceding year. He spent four days in reconnoitring, and then he sent his brother, with part of the fleet, to make a feint upon New York, and occupy Washington's attention. Before daybreak of the 27th, under the cover of this feint, the British advance had been nearly completed. General Grant, with the Highland regiments, advanced along the coast road, where the American outposts were held by William Alexander of New Jersey, commonly known as Lord Stirling, from a lapsed Scotch earldom to which he had claimed the title. The Hessians, under General von Heister, proceeded along the Bedford and Flatbush roads,

Battle of Long Island, Aug. 27, 1776.

which were defended by Sullivan; while more than half of the army, under Howe in person, accompanied by Clinton, Percy, and Cornwallis, accomplished a long night march by the Jamaica road, in order to take the Americans in flank. This long flanking march was completed in perfect secrecy because the people of the neighbourhood were in sympathy with the British, and it encountered no obstacles because the American force was simply incapable of covering so much territory. The divisions of Stirling and Sullivan contained the 5,000 men which were all that Putnam could afford to send forward from his works. A patrol which watched the Jamaica road was captured early in the morning, but it would not in any case have been possible to send any force there which could materially have hindered the British advance. Overwhelming superiority in numbers enabled the British to go where they pleased, and the battle was already virtually won when they appeared on the Jamaica road in the rear of the village of Bedford. Scarcely had the fight begun on the crest of the hill between Sullivan and the Hessians in his front when he found himself assaulted in the rear. Thrown into confusion, and driven back and forth through the woods between two galling fires, his division was quickly routed, and nearly all were taken prisoners, including the general himself. On the coast road the fight between Stirling and Grant was the first in which Americans had ever met British troops in open field and in regular line of battle. Against the sturdy Highland regiments Stirling held his ground gallantly

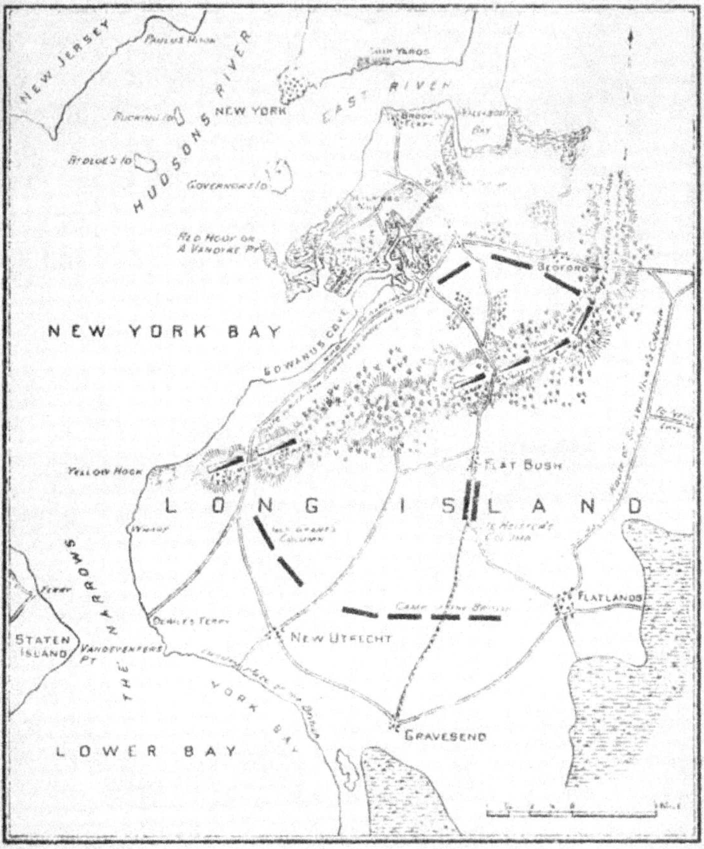

BATTLE OF LONG ISLAND

August 27, 1776

for four hours, until he was in turn assaulted in the rear by Lord Cornwallis, after the rout of Sullivan. It now became, with Stirling, simply a question of saving his division from capture, and after a desperate fight this end was accomplished, and the men got back to Brooklyn Heights, though the brave Stirling himself was taken prisoner. In this noble struggle the highest honours were won by the brigade of Maryland men commanded by Smallwood, and throughout the war we shall find this honourable distinction of Maryland for the personal gallantry of her troops fully maintained, until in the last pitched battle, at Eutaw Springs, we see them driving the finest infantry of England at the point of the bayonet.

The defeat of Sullivan and Stirling enabled Howe to bring up his whole army in front of the works at Brooklyn Heights toward the close of the day. To complete the victory it would be necessary to storm these works, but Howe's men were tired with marching, if not with fighting, and so the incident known as the battle of Long Island came to an end. A swift ship was at once dispatched to England with the news of the victory, which were somewhat highly coloured. It was for a while supposed that there had been a terrible slaughter, but careful research has shown that this was not the case. About 400 had been killed and wounded on each side, and this loss had been incurred mainly in the fight between Stirling and Grant. On other parts of the field the British triumph had consisted chiefly in the scooping up of prisoners, of whom at least 1,000 were taken. The stories of a wholesale butchery by the Hessians which once were current have been completely disproved. Washington gave a detailed account of the affair a few days afterward, and the most careful investigation has shown that he was correct in every particular. But to the American public the blow was none the less terrible, while in England the exultation served as an offset to the chagrin felt after the loss of Boston and the defeat at Fort Moultrie, and it was naturally long before facts could be seen in their true proportions.

Heavy as was the blow, however, General Howe's object was still but half attained. He had neither captured nor destroyed the American forces on Long Island, but had only driven them into their works. He was still confronted by 8,000 men on Brooklyn Heights, and the problem was how to dislodge them. In the evening Washington came over from New York, and made everything ready to resist a storm. To this end, on the next day, he brought over reinforcements, raising his total force within the works to 10,000 men. Under such circumstances, if the British had attempted a storm they would probably have been repulsed with great slaughter. But Howe had not forgotten Bunker Hill, and he thought it best to proceed by way of siege. As soon as Washington perceived this intention of his adversary, he saw that he must withdraw his army. He would have courted a storm, in which he was almost sure to be victorious, but he shrank from a siege, in which he was quite sure to lose his whole force. The British troops now invested him in a

Howe prepares to besiege the Heights;

semicircle, and their ships might at any moment close in behind and cut off his only retreat. Accordingly, sending trusty messengers across the river, Washington collected every sloop, yacht, fishing-smack, yawl, scow, or row-boat that could be found in either water from the Battery to King's Bridge or Hell-Gate; and after nightfall of the 29th, these craft were all assembled at the Brooklyn ferry, and wisely manned by the fishermen of Marblehead and Gloucester from Glover's Essex regiment, experts, every one of them, whether at oar or sail. All through the night the American troops were ferried across the broad river, as quietly as possible and in excellent order, while Washington superintended the details of the embarkation, and was himself the last man to leave the ground. At seven o'clock in the morning the whole American army had landed on the New York side, and had brought with them all their cannon, small arms, ammunition, tools, and horses, and all their larder besides, so that when the bewildered British climbed into the empty works they did not find so much as a biscuit or a glass of rum wherewith to console themselves.

but Washington slips away with his army.

This retreat has always been regarded as one of the most brilliant incidents in Washington's career, and it would certainly be hard to find a more striking example of vigilance. Had Washington allowed himself to be cooped up on Brooklyn Heights he would have been forced to surrender; and whatever was left of the war would have been

His vigilance robbed the British of the most golden opportunity ever afforded them.

a game played without queen, rook, or bishop. For this very reason it is hardly creditable to Howe that he should have let his adversary get away so easily. At daybreak, indeed, the Americans had been remarkably favoured by the sudden rise of a fog which covered the East river, but during the night the moon had shone brightly, and one can only wonder that the multitudinous plash of oars and the unavoidable murmur of ten thousand men embarking, with their heavy guns and stores, should not have attracted the attention of some wakeful sentinel, either on shore or on the fleet. A storming party of British, at the right moment, would at least have disturbed the proceedings. So rare a chance of ending the war at a blow was never again to be offered to the British commanders. Washington now stationed the bulk of his army along the line of the Harlem river, leaving a strong detachment in the city under Putnam; and presently, with the same extraordinary skill which he had just displayed in sending boats under the very eyes of the fleet, he withdrew Colonel Prescott and his troops from their exposed position on Governor's Island, which there was no longer any reason for holding.

Hoping that the stroke just given by the British sword might have weakened the obstinacy of the Americans, Lord Howe again had recourse to the olive-branch. The captured General Sullivan was sent to Congress to hold out hopes that Lord Howe would use his influence to get all the obnoxious acts of Parliament repealed, only he would first like to confer with some of the members of Congress informally and as with mere private gentlemen. A lively debate ensued upon this proposal, in which some saw an insult to Congress, while all quite needlessly suspected treachery. John Adams, about whom there was so much less of the *suaviter in modo* than of the *fortiter in re*, alluded to Sullivan, quite unjustly, as a "decoy duck," who had better have been shot in the battle than employed on such a business. It was finally voted that no proposals of peace from Great Britain should receive notice, unless they should be conveyed in writing, and should explicitly recognize Congress as the legal representative of the American States. For this once, however, out of personal regard for Lord Howe, and that nothing might be disdained which really looked toward a peaceful settlement, they would send a committee to Staten Island to confer with his lordship, who might regard this committee in whatever light he pleased. In this shrewd, half-humorous method of getting rid of the diplomatic difficulty, one is forcibly reminded of President Lincoln's famous proclamation addressed "To whom it may concern." The committee, consisting of Franklin, Rutledge, and John Adams, were hospitably entertained by Lord Howe, but their conference came to nothing, because the Americans now demanded a recognition of their independence as a condition which must precede all negotiation. There is no doubt that Lord Howe, who was a warm friend to the Americans and an energetic opponent of the king's policy, was bitterly grieved at this result. As a last resort he published a

The conference at Staten Island, Sept. 11.

proclamation announcing the intention of the British government to reconsider the various acts and instructions by which the Americans had been annoyed, and appealing to all right-minded people to decide for themselves whether it were not wise to rely on a solemn promise like this, rather than commit themselves to the dangerous chances of an unequal and unrighteous war.

Four days after this futile interview General Howe took possession of New York. After the loss of Brooklyn Heights, Washington and Greene were already aware that the city could not be held. Its capture was very easily effected. Several ships-of-the-line ascended the Hudson as far as Bloomingdale, and the East river as far as Blackwell's Island; and while thus from either side these vessels swept the northern part of Manhattan with a searching fire, General Howe brought his army across from Brooklyn in boats and landed at Kipp's Bay, near the present site of East Thirty-Fourth Street. Washington came promptly down, with two New England brigades, to reinforce the men whom he had stationed at that point, and to hinder the landing of the enemy until Putnam should have time to evacuate the city. To Washington's wrath and disgust, these men were seized with panic, and suddenly turned and fled without firing a shot. Had Howe now thrown his men promptly forward across the line of Thirty-Fourth Street, he would have cut off Putnam's retreat from the city. But what the New England brigades failed to do a bright woman succeeded in accomplishing. When Howe

Howe takes the city of New York, Sept. 15;

had reached the spot known as Murray Hill,— now the centre of much brownstone magnificence in Park and Madison and Fifth avenues, at that time a noble country farmstead,—Mrs. Lindley Murray, mother of the famous grammarian, well knowing the easy temper of the British commander, sent out a servant to invite him to stop and take luncheon. A general halt was ordered; and while Howe and his officers were gracefully entertained for more than two hours by their accomplished and subtle hostess, Putnam hastily marched his 4,000 men up the shore of the Hudson, until, passing Bloomingdale, he touched the right wing of the main army, and was safe, though his tents, blankets, and heavy guns had been left behind. The American lines now extended from the mouth of Harlem river across the island, and on the following day the British attempted to break through their centre at Harlem Heights; but the attack was repulsed, with a loss of sixty Americans and three hundred British, and the lines just formed remained, with very little change, for nearly four weeks.

but Mrs. Lindley Murray saves the garrison.

Attack upon Harlem Heights, Sept. 16.

General Howe had thus got possession of the city of New York, but the conquest availed him little so long as the American army stood across the island, in the attitude of blockading him. If this campaign was to decide the war, as the ministry hoped, nothing short of the capture or dispersal of Washington's army would suffice. But the problem was now much harder than it had been at Brooklyn.

The new problem before Howe.

For as the land above Manhattan Island widens rapidly to the north and east, it would not be easy to hem Washington in by sending forces to his rear. As soon as he should find his position imperilled, he would possess the shorter line by which to draw his battalions together and force an escape, and so the event proved. Still, with Howe's superior force and with his fleet, if he could get up the Hudson to the rear of the American right, and at the same time land troops from the Sound in the rear of the American left, it was possible that Washington might be compelled to surrender. There was nothing to bar Howe's passage up the East river to the Sound; but at the northern extremity of Manhattan Island the ascent of the Hudson was guarded on the east by Fort Washington, under command of Putnam, and on the west by Fort Lee, standing on the summit of the lofty cliffs known as the Palisades, and commanded by Greene. It was still doubtful, however, whether these two strongholds could effectually bar the ascent of so broad a river, and for further security Putnam undertook to place obstructions in the bed of the stream itself. Both the Continental Congress and the State Convention of New York were extremely unwilling that these two fortresses should in any event be given up, for in no case must the Hudson river be abandoned. Putnam and Greene thought that the forts could be held, but by the 9th of October it was proved that they could not bar the passage of the river, for on that day two frigates ran safely between them, and captured some small American craft a short distance above.

This point having been ascertained, General Howe, on the 12th, leaving Percy in command before Harlem Heights, moved the greater part of his army nine miles up the East river to Throg's Neck, a peninsula in the Sound, separated from the mainland by a narrow creek and a marsh that was overflowed at high tide. By landing here suddenly, Howe hoped to get in Washington's rear and cut him off from his base of supply in Connecticut. But Washington had foreseen the move and forestalled it. When Howe arrived at Throg's Neck, he found the bridge over the creek destroyed, and the main shore occupied by a force which it would be dangerous to try to dislodge by wading across the marsh. While Howe was thus detained six days on the peninsula Washington moved his base to White Plains, and concentrated his whole army at that point, abandoning everything on Manhattan Island except Fort Washington. Sullivan, Stirling, and Morgan, who had just been exchanged, now rejoined the army, and Lee also arrived from South Carolina.

Howe moves upon Throg's Neck, but Washington changes base.

By this movement to White Plains, Washington had foiled Howe's attempt to get in his rear, and the British general decided to try the effect of an attack in front. On the 28th of October he succeeded in storming an outpost at Chatterton Hill, losing 229 lives, while the Americans lost 140. But this affair, which is sometimes known as the battle of White Plains, seems to have discouraged Howe. Before renewing the attack he waited three days,

Baffled at White Plains, Howe tries a new plan.

thinking perhaps of Bunker Hill; and on the last night of October, Washington fell back upon North Castle, where he took a position so strong that it was useless to think of assailing him. Howe then changed his plans entirely, and moved down the east bank of the Hudson to Dobb's Ferry, whence he could either attack Fort Washington or cross into New Jersey and advance upon Philadelphia, the "rebel capital." The purpose of this change was to entice Washington from his unassailable position.

To meet this new movement, Washington threw his advance of 5,000 men, under Putnam, into New Jersey, where they encamped near Hackensack; he sent Heath up to Peekskill, with 3,000 men, to guard the entrance to the Highlands; and he left Lee at North Castle, with 7,000 men, and ordered him to cöoperate with him promptly in whatever direction, as soon as the nature of Howe's plans should become apparent. As Forts Washington and Lee detained a large force in garrison, while they had shown themselves unable to prevent ships from passing up the river, there was no longer any use in holding them. Nay, they had now become dangerous, as traps in which the garrisons and stores might be suddenly surrounded and captured. Washington accordingly resolved to evacuate them both, while, to allay the fears of Congress in the event of a descent from Canada, he ordered Heath to fortify the much more important position at West Point.

Washington's orders in view of the emergency.

Had Washington's orders been obeyed and his plans carried out, history might still have recorded

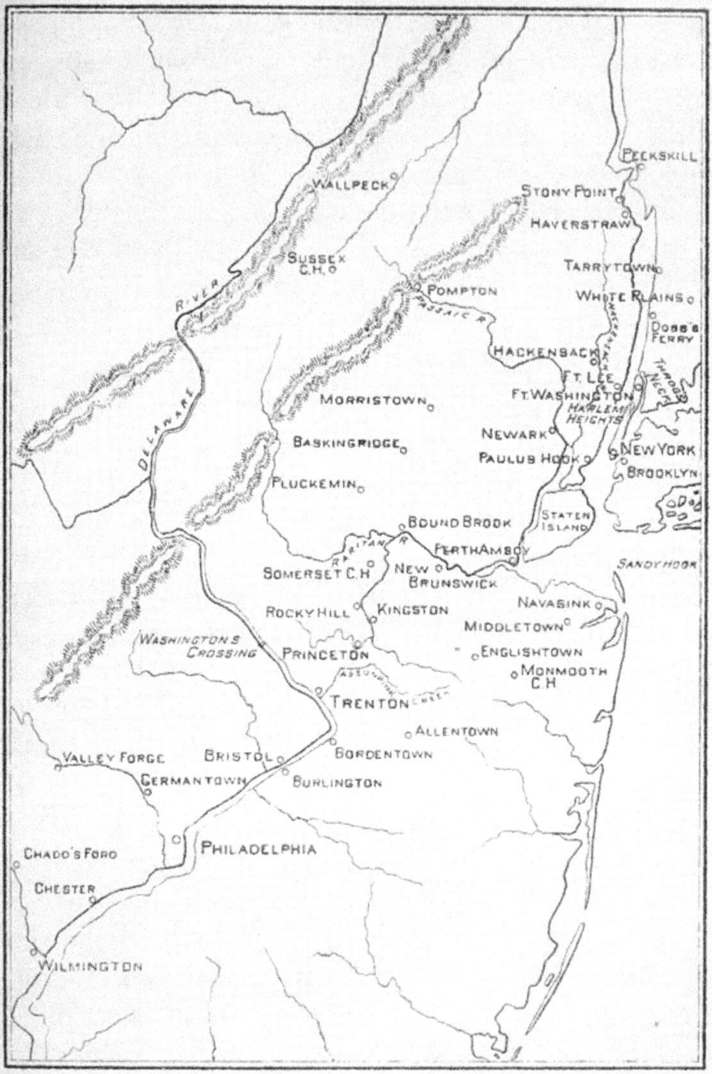

OPERATIONS IN NEW YORK AND NEW JERSEY

1776–77

a retreat through "the Jerseys," but how different a retreat from that which was now about to take place! The officious interference of Congress, a venial error of judgment on the part of Greene, and gross insubordination on the part of Lee, occurring all together at this critical moment, brought about the greatest disaster of the war, and came within an ace of overwhelming the American cause in total and irretrievable ruin. Washington instructed Greene, who now commanded both fortresses, to withdraw the garrison and stores from Fort Washington, and to make arrangements for evacuating Fort Lee also. At the same time he did not give a positive order, but left the matter somewhat within Greene's discretion, in case military circumstances of an unforeseen kind should arise. Then, while Washington had gone up to reconnoitre the site for the new fortress at West Point, there came a special order from Congress that Fort Washington should not be abandoned save under direst extremity. If Greene had thoroughly grasped Washington's view of the case, he would have disregarded this conditional order, for there could hardly be a worse extremity than that which the sudden capture of the fortress would entail. But Greene's mind was not quite clear; he believed that the fort could be held, and he did not like to take the responsibility of disregarding a message from Congress. In this dilemma he did the worst thing possible: he reinforced the doomed garrison, and awaited Washington's return.

Congress meddles with the situation and muddles it.

When the commander-in-chief returned, on the 14th, he learned with dismay that nothing had been done. But it was now too late to mend matters, for that very night several British vessels passed up between the forts, and the next day Howe appeared before Fort Washington with an overwhelming force, and told Colonel Magaw, the officer in charge, that if he did not immediately surrender the whole garrison would be put to the sword. Magaw replied that if Howe wanted his fort he must come and take it. On the 16th, after a sharp struggle, in which the Americans fought with desperate gallantry, though they were outnumbered more than five to one, the works were carried, and the whole garrison was captured. The victory cost the British more than 500 men in killed and wounded. The Americans, fighting behind their works, lost but 150; but they surrendered 3,000 of the best troops in their half-trained army, together with an immense quantity of artillery and small arms. It was not in General Howe's kindly nature to carry out his savage threat of the day before; but some of the Hessians, maddened with the stubborn resistance they had encountered, began murdering their prisoners in cold blood, until they were sharply called to order. From Fort Lee, on the opposite bank of the river, Washington surveyed this woful surrender with his usual iron composure; but when it came to seeing his brave men thrown down and stabbed to death by the Hessian bayonets, his overwrought heart could bear it no longer, and he cried and sobbed like a child.

Howe takes Fort Washington by storm, Nov. 16.

This capture of the garrison of Fort Washington was one of the most crushing blows that befell the American arms during the whole course of the war. Washington's campaign seemed now likely to be converted into a mere flight, and a terrible gloom overspread the whole country. The disaster was primarily due to the interference of Congress. It might have been averted by prompt and decisive action on the part of Greene. But Washington, whose clear judgment made due allowance for all the circumstances, never for a moment cast any blame upon his subordinate. The lesson was never forgotten by Greene, whose intelligence was of that high order which may indeed make a first mistake, but never makes a second. The friendship between the two generals became warmer than ever. Washington, by a sympathetic instinct, had divined from the outset the military genius that was by and by to prove scarcely inferior to his own.

Washington and Greene.

Yet worse remained behind. Washington had but 6,000 men on the Jersey side of the river, and it was now high time for Lee to come over from North Castle and join him, with the force of 7,000 that had been left under his command. On the 17th, Washington sent a positive order for him to cross the river at once; but Lee dissembled, pretended to regard the order in the light of mere advice, and stayed where he was. He occupied an impregnable position: why should he leave it, and imperil a force with which he might accomplish something memorable on his own account? By the resignation of

Outrageous conduct of Charles Lee.

General Ward, Lee had become the senior major-general of the Continental army, and in the event of disaster to Washington he would almost certainly become commander-in-chief. He had returned from South Carolina more arrogant and loud-voiced than ever. The northern people knew little of Moultrie, while they supposed Lee to be a great military light; and the charlatan accordingly got the whole credit of the victory, which, if his precious advice had been taken, would never have been won. Lee was called the hero of Charleston, and people began to contrast the victory of Sullivan's Island with the recent defeats, and to draw conclusions very disparaging to Washington. From the beginning Lee had felt personally aggrieved at not being appointed to the chief command, and now he seemed to see a fair chance of ruining his hated rival. Should he come to the head of the army in a moment of dire disaster to the Americans, it would be so much the better, for it would be likely to open negotiations with Lord Howe, and Lee loved to chaffer and intrigue much better than to fight. So he spent his time in endeavouring, by insidious letters and lying whispers, to nourish the feeling of disaffection toward Washington, while he refused to send a single regiment to his assistance. Thus, through the villainy of this traitor in the camp, Washington actually lost more men, so far as their present use was concerned at this most critical moment, than he had been deprived of by all the blows which the enemy had dealt him since the beginning of the campaign.

On the night of the 19th, Howe threw 5,000 men across the river, about five miles above Fort Lee, and with this force Lord Cornwallis marched rapidly down upon that stronghold. The place had become untenable, and it was with some difficulty that a repetition of the catastrophe of Fort Washington was avoided. Greene had barely time, with his 2,000 men, to gain the bridge over the Hackensack and join the main army, leaving behind all his cannon, tents, blankets, and eatables. The position now occupied by the main army, between the Hackensack and Passaic rivers, was an unsafe one, in view of the great superiority of the enemy in numbers. A strong British force, coming down upon Washington from the north, might compel him to surrender or to fight at a great disadvantage. To avoid this danger, on the 21st, he crossed the Passaic and marched southwestward to Newark, where he stayed five days; and every day he sent a messenger to Lee, urging him to make all possible haste in bringing over his half of the army, that they might be able to confront the enemy on something like equal terms. Nothing could have been more explicit or more peremptory than Washington's orders; but Lee affected to misunderstand them, sent excuses, raised objections, paltered, argued, prevaricated, and lied, and so contrived to stay where he was until the first of December. To Washington he pretended that his moving was beset by "obstacles," the nature of which he would explain as soon as they should meet. But to James Bowdoin, president of the executive council

Greene barely escapes from Fort Lee, Nov. 20.

of Massachusetts, he wrote at the same time declaring that his own army and that under Washington "must rest each on its own bottom." He assumed command over Heath, who had been left to guard the Highlands, and ordered him to send 2,000 troops to reinforce the main army; but that officer very properly refused to depart from the instructions which the commander-in-chief had left with him. To various members of Congress Lee told the falsehood that if *his* advice had only been heeded, Fort Washington would have been evacuated ere it was too late; and he wrote to Dr. Rush, wondering whether any of the members of Congress had ever studied Roman history, and suggesting that he might do great things if he could only be made Dictator for one week.

Lee intrigues against Washington.

Meanwhile Washington, unable to risk a battle, was rapidly retreating through New Jersey. On the 28th of November Cornwallis advanced upon Newark, and Washington fell back upon New Brunswick. On the first of December, as Cornwallis reached the latter place, Washington broke down the bridge over the Raritan, and continued his retreat to Princeton. The terms of service for which his troops had been enlisted were now beginning to expire, and so great was the discouragement wrought by the accumulation of disasters which had befallen the army since the battle of Long Island that many of the soldiers lost heart in their work. Homesickness began to prevail, especially among the New England troops, and as their terms ex-

Washington retreats into Pennsylvania.

pired it was difficult to persuade them to reënlist. Under these circumstances the army dwindled fast, until, by the time he reached Princeton, Washington had but 3,000 men remaining at his disposal. The only thing to be done was to put the broad stream of the Delaware between himself and the enemy, and this he accomplished by the 8th, carrying over all his guns and stores, and seizing or destroying every boat that could be found on that great river for many miles in either direction. When the British arrived, on the evening of the same day, they found it impossible to cross. Cornwallis was eager to collect a flotilla of boats as soon as practicable, and push on to Philadelphia, but Howe, who had just joined him, thought it hardly worth while to take so much trouble, as the river would be sure to freeze over before many days. So the army was posted — with front somewhat too far extended — along the east bank, with its centre at Trenton, under Colonel Rall; and while they waited for that "snap" of intensely cold weather, which in this climate seldom fails to come on within a few days of Christmas, Howe and Cornwallis both went back to New York.

Meanwhile, on the 2d of December, Lee had at last crossed the Hudson with a force diminished to 4,000 men, and had proceeded by slow marches as far as Morristown. Further reinforcements were at hand. General Schuyler, in command of the army which had retreated the last summer from Canada, was guarding the forts on Lake Champlain; and as these appeared to be safe for the present, he detached seven regiments to go to the aid of Washington. As soon as Lee heard of the arrival of three of these regiments at Peekskill, he ordered them to join him at Morristown. As the other four, under General Gates, were making their way through northern New Jersey, doubts arose as to where they should find Washington in the course of his swift retreat. Gates sent his aid, Major Wilkinson, forward for instructions, and he, learning that Washington had withdrawn into Pennsylvania, reported to Lee at Morristown, as second in command.

Reinforcements come from Schuyler.

Lee had left his army in charge of Sullivan, and had foolishly taken up his quarters at an unguarded tavern about four miles from the town, where Wilkinson found him in bed on the morning of the 13th. After breakfast Lee wrote a confidential letter to Gates, as to a kindred spirit from whom he might expect to get sympathy. Terrible had been the consequences of the disaster at Fort Washington. "There never was so damned a stroke," said the letter. "*Entre nous*, a certain great man is most damnably deficient. He has thrown me into a situation where I have my choice of difficulties. If I stay in this province I risk myself and army, and if I do not stay the province is lost forever. . . . Our counsels have been weak to the last degree. As to yourself, if you think you can be in time to aid the general, I would have you by all means go. You will at least save your army. . . . Adieu, my dear friend. God bless you." Hardly had he signed his name to this scan-

Fortunately for the Americans, the British capture Charles Lee, Dec. 13.

dalous document when Wilkinson, who was standing at the window, exclaimed that the British were upon them. Sure enough. A Tory in the neighbourhood, discerning the golden opportunity, had galloped eighteen miles to the British lines, and returned with a party of thirty dragoons, who surrounded the house and captured the vainglorious schemer before he had time to collect his senses. Bareheaded, and dressed only in a flannel gown and slippers, he was mounted on Wilkinson's horse, which stood waiting at the door, and was carried off, amid much mirth and exultation, to the British camp. Crest-fallen and bewildered, he expressed a craven hope that his life might be spared, but was playfully reminded that he would very likely be summarily dealt with as a deserter from the British army; and with this scant comfort he was fain to content himself for some weeks to come.

The capture of General Lee was reckoned by the people as one more in the list of dire catastrophes which made the present season the darkest moment in the whole course of the war. Had they known all that we know now, they would have seen that the army was well rid of a worthless mischief-maker, while the history of the war had gained a curiously picturesque episode. Apart from this incident there was cause enough for the gloom which now overspread the whole country. Washington had been forced to seek shelter behind the Delaware with a handful of men, whose terms of service were soon to expire, and another fortnight might easily witness the utter dispersal of this poor little army. At Philadelphia, where

Putnam was now in command, there was a general panic, and people began hiding their valuables and moving their wives and children out into the country. Congress took fright, and retired to Baltimore. At the beginning of December, Lord Howe and his brother had issued a proclamation offering pardon and protection to all citizens who within sixty days should take the oath of allegiance to the British Crown; and in the course of ten days nearly three thousand persons, many of them wealthy and of high standing in society, had availed themselves of this promise. The British soldiers and the Tories considered the contest virtually ended. General Howe was compared with Cæsar, who came, and saw, and conquered. For his brilliant successes he had been made a Knight Commander of the Bath, and New York was to become the scene of merry Christmas festivities on the occasion of his receiving the famous red ribbon. In his confidence that Washington's strength was quite exhausted, he detached a considerable force from the army in New Jersey, and sent it, under Lord Percy, to take possession of Newport as a convenient station for British ships entering the Sound. Donop and Rall with their Hessians and Grant with his hardy Scotchmen would now quite suffice to destroy the remnant of Washington's army; and Cornwallis accordingly packed his portmanteaus and sent them aboard ship, intending to sail for England as soon as the fumes of the Christmas punch should be duly slept off.

The times that tried men's souls.

Well might Thomas Paine declare, in the first of the series of pamphlets entitled "The Crisis," which he now began to publish, that "these are the times that try men's souls." But in the midst of the general despondency there were a few brave hearts that had not yet begun to despair, and the bravest of these was Washington's. At this awful moment the whole future of America, and of all that America signifies to the world, rested upon that single Titanic will. Cruel defeat and yet more cruel treachery, enough to have crushed the strongest, could not crush Washington. All the lion in him was aroused, and his powerful nature was aglow with passionate resolve. His keen eye already saw the elements of weakness in Howe's too careless disposition of his forces on the east bank of the Delaware, and he had planned for his antagonist such a Christmas greeting as he little expected. Just at this moment Washington was opportunely reinforced by Sullivan and Gates, with the troops lately under Lee's command; and with his little army thus raised to 6,000 men, he meditated such a stroke as might revive the drooping spirits of his countrymen, and confound the enemy in the very moment of his fancied triumph.

Washington prepares to strike back.

Washington's plan was, by a sudden attack, to overwhelm the British centre at Trenton, and thus force the army to retreat upon New York. The Delaware was to be crossed in three divisions. The right wing, of 2,000 men, under Gates, was to attack Count Donop at Burlington; Ewing, with the centre, was to cross directly opposite Trenton; while Washington himself, with the left wing, was to cross nine miles above, and march down upon

Trenton from the north. On Christmas Day all was ready, but the beginnings of the enterprise were not auspicious. Gates, who preferred to go and intrigue in Congress, succeeded in begging off, and started for Baltimore. Cadwalader, who took his place, tried hard to get his men and artillery across the river, but was baffled by the huge masses of floating ice, and reluctantly gave up the attempt. Ewing was so discouraged that he did not even try to cross, and both officers took it for granted that Washington must be foiled in like manner. But Washington was desperately in earnest; and although at sunset, just as he had reached his crossing-place, he was informed by special messenger of the failure of Ewing and Cadwalader, he determined to go on and make the attack with the 2,500 men whom he had with him. The great blocks of ice, borne swiftly along by the powerful current, made the passage extremely dangerous, but Glover, with his skilful fishermen of Marblehead, succeeded in ferrying the little army across without the loss of a man or a gun. More than ten hours were consumed in the passage, and then there was a march of nine miles to be made in a blinding storm of snow and sleet. They pushed rapidly on in two columns, led by Greene and Sullivan respectively, drove in the enemy's pickets at the point of the bayonet, and entered the town by different roads soon after sunrise. Washington's guns were at once planted so as to sweep the streets, and after Colonel Rall and seventeen of his men had been slain, the whole body of Hessians, 1,000 in num-

He crosses the Delaware;

and pierces the British centre at Trenton, Dec. 26.

ber, surrendered at discretion. Of the Americans, two were frozen to death on the march, and two were killed in the action. By noon of the next day Cadwalader had crossed the river to Burlington, but no sooner had Donop heard what had happened at Trenton than he retreated by a circuitous route to Princeton, leaving behind all his sick and wounded soldiers, and all his heavy arms and baggage. Washington recrossed into Pennsylvania with his prisoners, but again advanced, and occupied Trenton on the 29th.

Cornwallis comes up to retrieve the disaster;

When the news of the catastrophe reached New York, the holiday feasting was rudely disturbed. Instead of embarking for England, Cornwallis rode post-haste to Princeton, where he found Donop throwing up earthworks. On the morning of January 2d Cornwallis advanced, with 8,000 men, upon Trenton, but his march was slow and painful. He was exposed during most of the day to a galling fire from parties of riflemen hidden in the woods by the roadside, and Greene, with a force of 600 men and two field-pieces, contrived so to harass and delay him that he did not reach Trenton till late in the afternoon. By that time Washington had withdrawn his whole force beyond the Assunpink, a small river which flows into the Delaware just south of Trenton, and had guarded the bridge and the fords by batteries admirably placed. The British made several attempts to cross, but were repulsed with some slaughter; and as their day's work had sorely fatigued them, Cornwallis thought best to wait until to-morrow, while he sent his messenger post-haste back to Princeton to bring up a force of nearly 2,000 men which he had left behind there. With this added strength he felt sure that he could force the passage of the stream above the American position, when by turning Washington's right flank he could fold him back against the Delaware, and thus compel him to surrender. Cornwallis accordingly went to bed in high spirits. "At last we have run down the old fox," said he, "and we will bag him in the morning."

and thinks he has run down the "old fox."

The situation was indeed a very dangerous one; but when the British general called his antagonist an old fox, he did him no more than justice. In its union of slyness with audacity, the movement which Washington now executed strongly reminds one of "Stonewall" Jackson. He understood perfectly well what Cornwallis intended to do; but he knew at the same time that detachments of the British army must have been left behind at Princeton and New Brunswick to guard the stores. From the size of the army before him he rightly judged that these rear detachments must be too small to withstand his own force. By overwhelming one or both of them, he could compel Cornwallis to retreat upon New York, while he himself might take up an impregnable position on the heights about Morristown, from which he might threaten the British line and hold their whole army in check, — a most brilliant and daring scheme for a commander to entertain while in such a perilous position as Washington was that night! But the manner

But Washington prepares a checkmate;

in which he began by extricating himself was not the least brilliant part of the manœuvre. All night long the American camp-fires were kept burning brightly, and small parties were busily engaged in throwing up intrenchments so near the Assunpink that the British sentinels could plainly hear the murmur of their voices and the thud of the spade and pickaxe. While this was going on, the whole American army marched swiftly up the south bank of the little stream, passed around Cornwallis's left wing to his rear, and gained the road to Princeton. Toward sunrise, as the British detachment was coming down the road from Princeton to Trenton, in obedience to Cornwallis's order, its van, under Colonel Mawhood, met the foremost column of Americans approaching, under General Mercer. As he caught sight of the Americans, Mawhood thought that they must be a party of fugitives, and hastened to intercept them; but he was soon undeceived. The Americans attacked with vigour, and a sharp fight was sustained, with varying fortunes, until Mercer was pierced by a bayonet, and his men began to fall back in some confusion. Just at this critical moment Washington came galloping upon the field and rallied the troops, and as the entire forces on both sides had now come up the fight became general. In a few minutes the British were routed and their line was cut in two; one half fleeing toward Trenton, the other half toward New Brunswick. There was little slaughter, as the whole fight did not occupy more than twenty minutes. The British lost about 200 in

and again severs the British line at Princeton, Jan. 3.

killed and wounded, with 300 prisoners and their cannon; the American loss was less than 100.

Shortly before sunrise, the men who had been left in the camp on the Assunpink to feed the fires and make a noise beat a hasty retreat, and found their way to Princeton by circuitous paths. When Cornwallis got up, he could hardly believe his eyes. Here was nothing before him but an empty camp: the American army had vanished, and whither it had gone he could not imagine. But his perplexity was soon relieved by the booming of distant cannon on the Princeton road, and the game which the "old fox" had played him all at once became apparent. Nothing was to be done but to retreat upon New Brunswick with all possible haste, and save the stores there.

General retreat of the British toward New York.

His road led back through Princeton, and from Mawhood's fugitives he soon heard the story of the morning's disaster. His march was hindered by various impediments. A thaw had set in, so that the little streams had swelled into roaring torrents, difficult to ford, and the American army, which had passed over the road before daybreak, had not forgotten to destroy the bridges. By the time that Cornwallis and his men reached Princeton, wet and weary, the Americans had already left it, but they had not gone on to New Brunswick. Washington had hoped to seize the stores there, but the distance was eighteen miles, his men were wretchedly shod and too tired to march rapidly, and it would not be prudent to risk a general engagement when his main purpose could be secured without one. For these reasons, Washington turned northward to the heights of Morristown, while Cornwallis continued his retreat to New Brunswick. A few days later, Putnam advanced from Philadelphia and occupied Princeton, thus forming the right wing of the American army, of which the main body lay at Morristown, while Heath's division on the Hudson constituted the left wing. Various cantonments were established along this long line. On the 5th, George Clinton, coming down from Peekskill, drove the British out of Hackensack and occupied it, while on the same day a detachment of German mercenaries at Springfield was routed by a body of militia. Elizabethtown was then taken by General Maxwell, whereupon the British retired from Newark.

Thus in a brief campaign of three weeks Washington had rallied the fragments of a defeated and broken army, fought two successful battles, taken nearly 2,000 prisoners, and recovered the state of New Jersey. He had cancelled the disastrous effects of Lee's treachery, and replaced things apparently in the condition in which the fall of Fort Washington had left them. Really he had done much more than this, for by assuming the offensive and winning victories through sheer force of genius, he had completely turned the tide of popular feeling. The British generals began to be afraid of him, while on the other hand his army began to grow by the accession of fresh recruits. In New Jersey, the enemy retained nothing but New Brunswick, Amboy, and Paulus Hook.

The tables completely turned.

On the 25th of January Washington issued a

proclamation declaring that all persons who had accepted Lord Howe's offer of protection must either retire within the British lines, or come forward and take the oath of allegiance to the United States. Many narrow-minded people, who did not look with favour upon a close federation of the states, commented severely upon the form of this proclamation: it was too national, they said. But it proved effective. However lukewarm may have been the interest which many of the Jersey people felt in the war when their soil was first invaded, the conduct of the British troops had been such that every one now looked upon them as foreign enemies. They had not only foraged indiscriminately upon friend and foe, but they had set fire to farmhouses, murdered peaceful citizens, and violated women. The wrath of the people, kindled by such outrages, had waxed so hot that it was not safe for the British to stir beyond their narrow lines except in considerable force. Their foraging parties were waylaid and cut off by bands of indignant yeomanry, and so sorely were they harassed in their advanced position at New Brunswick that they often suffered from want of food. Many of the German mercenaries, caring nothing for the cause in which they had been forcibly enlisted, began deserting; and in this they were encouraged by Congress, which issued a manifesto in German, making a liberal offer of land to any foreign soldier who should leave the British service. This little document was inclosed in the wrappers in which packages of tobacco were sold, and every now and then some canny smoker accepted the offer.

Washington's position at Morristown was so strong that there was no hope of dislodging him, and the snow-blocked roads made the difficulties of a winter campaign so great that Howe thought best to wait for warm weather before doing anything more. While the British arms were thus held in check, the friends of America, both in England and on the continent of Europe, were greatly encouraged. From this moment Washington was regarded in Europe as a firstrate general. Military critics who were capable of understanding his movements compared his brilliant achievements with his slender resources, and discovered in him genius of a high order. Men began to call him "the American Fabius;" and this epithet was so pleasing to his fellow-countrymen, in that pedantic age, that it clung to him for the rest of his life, and was repeated in newspapers and speeches and pamphlets with wearisome iteration. Yet there was something more than Fabian in Washington's generalship. For wariness he has never been surpassed; yet, as Colonel Stedman observed, in his excellent contemporary history of the war, the most remarkable thing about Washington was his courage. It would be hard indeed to find more striking examples of audacity than he exhibited at Trenton and Princeton. Lord Cornwallis was no mean antagonist, and no one was a better judge of what a commander might be expected to do with a given stock of resources. His surprise at the Assunpink was so great that he never got over it. After the surrender at Yorktown, it is said that his lord-

Washington's superb generalship.

ship expressed to Washington his generous admiration for the wonderful skill which had suddenly hurled an army four hundred miles, from the Hudson river to the James, with such precision and such deadly effect. "But after all," he added, "your excellency's achievements in New Jersey were such that nothing could surpass them." The man who had turned the tables on him at the Assunpink he could well believe to be capable of anything.

In England the effect of the campaign was very serious. Not long before, Edmund Burke had despondingly remarked that an army which was always obliged to refuse battle could never expel the invaders; but now the case wore a different aspect. Sir William Howe had not so much to show for his red ribbon, after all. He had taken New York, and dealt many heavy blows with his overwhelming force, unexpectedly aided by foul play on the American side; but as for crushing Washington and ending the war, he seemed farther from it than ever. It would take another campaign to do this, — perhaps many. Lord North, who had little heart for the war at any time, was discouraged, while the king and Lord George Germain were furious with disappointment. "It was that unhappy affair of Trenton," observed the latter, "that blasted our hopes."

In France the interest in American affairs grew rapidly. Louis XVI. had no love for Americans or for rebels, but revenge for the awful disasters of 1758 and 1759 was dear to the French heart. France felt toward England then as she feels toward Germany now, and so long ago as the time of the Stamp Act, Baron Kalb had been sent on a secret mission to America, to find out how the people regarded the British government. The policy of the French ministry was aided by the romantic sympathy for America which was felt in polite society. Never perhaps have the opinions current among fashionable ladies and gentlemen been so directly controlled by philosophers and scholars as in France during the latter half of the eighteenth century. Never perhaps have men of letters exercised such mighty influence over their contemporaries as Voltaire, with his noble enthusiasm for humanity, and Rousseau, with his startling political paradoxes, and the writers of the "Encyclopédie," with their revelations of new points of view in science and in history. To such men as these, and to such profound political thinkers as Montesquieu and Turgot, the preservation and extension of English liberty was the hope of the world; but they took little interest in the British crown or in the imperial supremacy of Parliament. All therefore sympathized with the Americans and urged on the policy which the court for selfish reasons was inclined to pursue. Vergennes, the astute minister of foreign affairs, had for some time been waiting for a convenient opportunity to take part in the struggle, but as yet he had contented himself with furnishing secret assistance. For more than a year he had been intriguing, through Beaumarchais, the famous author of "Figaro," with Arthur Lee (a brother of Richard Henry Lee), who had long served in London as agent for Virginia. Just before the Declaration of Independence Vergennes sent over a million dollars to aid the American cause. Soon afterwards Congress sent Silas Deane to Paris, and presently ordered Arthur Lee to join him there. In October Franklin was also sent over, and the three were appointed commissioners for making a treaty of alliance with the French court.

The arrival of Franklin was the occasion of great excitement in the fashionable world of Paris. By thinkers like Diderot and D'Alembert he was regarded as the embodiment of practical wisdom. To many he seemed to sum up in himself the excellences of the American cause, — justice, good sense, and moderation. Voltaire spoke quite unconsciously of the American army as "Franklin's troops." It was Turgot who said of him, in a line which is one of the finest modern specimens of epigrammatic Latin, "Eripuit cœlo fulmen, sceptrumque tyrannis." As symbolizing the liberty for which all France was yearning, he was greeted with a popular enthusiasm such as perhaps no Frenchman except Voltaire has ever called forth. As he passed along the streets, the shopkeepers rushed to their doors to catch a glimpse of him, while curious idlers crowded the sidewalk. The charm of his majestic and venerable figure seemed heightened by the republican simplicity of his plain brown coat, over the shoulders of which his long gray hair fell carelessly, innocent of queue or powder. His portrait was hung in the shop-windows and painted in miniature on the covers of snuff-boxes. Gentlemen wore "Franklin" hats, ladies' kid gloves were dyed of a "Franklin" hue, and *cotelettes à la Franklin* were served at fashionable dinners.

As the first fruits of Franklin's negotiations, the French government agreed to furnish two million livres a year, in quarterly instalments, to assist the American cause. Three ships, laden with military stores, were sent over to America: one was captured by a British cruiser, but the other two arrived safely. The Americans were allowed to fit out privateers in French ports, and even to bring in and sell their prizes there. Besides this a million livres were advanced to the commissioners on account of a quantity of tobacco which they agreed to send in exchange. Further than this France was not yet ready to go. The British ambassador had already begun to protest against the violation of neutrality involved in the departure of privateers, and France was not willing to run the risk of open war with England until it should become clear that the Americans would prove efficient allies. The king, moreover, sympathized with George III., and hated the philosophers whose opinions swayed the French people; and in order to accomplish anything in behalf of the Americans he had to be coaxed or bullied at every step.

But though the French government was not yet ready to send troops to America, volunteers were not wanting who cast in their lot with us through a purely disinterested enthusiasm. At a dinner party in Metz, the Marquis de Lafayette, then a boy of nineteen, heard the news from America, and instantly resolved to leave his pleasant home and offer his services to Washington. He fitted up a ship at his own expense, loaded it with military stores furnished by Beaumarchais, and set sail from Bordeaux on the 26th of April, taking with him Kalb and eleven other officers. While Marie Antoinette applauded his generous self-devotion, the king forbade him to go, but he disregarded the order. His young wife, whom he deemed it prudent to leave behind, he consoled with the thought that the future welfare of all mankind was at stake in the struggle for constitutional liberty which was going on in America, and that where he saw a chance to be useful it was his duty to go. The able Polish officers, Pulaski and Kosciusko had come some time before.

During the winter season at Morristown, Washington was busy in endeavouring to recruit and reorganize the army. Up to this time the military preparations of Congress had been made upon a ludicrously inadequate scale. There had been no serious attempt to create a regular army, but squads of militia had been enlisted for terms of three or six months, as if there were any likelihood of the war being ended within such a period. The rumour of Lord Howe's olive-branch policy may at first have had something to do with this, and even after the Declaration of Independence had made further temporizing impossible, there were many who expected Washington to perform miracles and thought that by some crushing blow the invaders might soon be brought to terms. But the events of the autumn had shown that the struggle was likely to prove long and desperate, and there could be no doubt as to the imperative need of a regular army. To provide such an army was, however, no easy task. The Continental Congress was little more than an advisory body of delegates, and it was questionable how far it could exercise authority except as regarded the specific points which the constituents of these delegates had in view when they chose them. Congress could only recommend to the different states to raise their respective quotas of men, and each state gave heed to such a request according to its ability or its inclination. All over the country there was then, as always, a deep-rooted prejudice against standing armies. Even to-day, with our population of sixty-five millions, a proposal to increase our regular army to fifty thousand men, for the more efficient police of the Indian districts in Arizona and Montana, has been greeted by the press with tirades about military despotism. A century ago this feeling was naturally much stronger than it is to-day. The presence of standing armies in this country had done much toward bringing on the Revolution; and it was not until it had become evident that we must either endure the king's regulars or have regulars of our own that the people could be made to adopt the latter alternative. Under the influence of these feelings, the state militias were enlisted for very short terms, each under its local officers, so that they resembled a group of little allied armies. Such methods were fatal to military discipline. Such soldiers as had remained in the army ever since it first gathered itself together on the day of Lexington had now begun to learn something of military discipline; but it was impossible to maintain it in the face of the much greater number who kept coming and going at intervals of three months. With such fluctuations in strength, moreover, it was difficult to carry out any series of military operations. The Christmas night when Washington crossed the Delaware was the most critical moment of his career; for the terms of service of the greater part of his little army expired on New Year's Day, and but for the success at Trenton, they would almost certainly have disbanded. But in the exultant mood begotten of this victory, they were persuaded to remain for some weeks longer, thus enabling Washington to recover the state of New Jersey. So low had the public credit sunk, at this season of disaster, that Washington pledged his private fortune for the payment of these men, in case Congress should be found wanting; and his example was followed by the gallant John Stark and other officers. Except for the sums raised by Robert Morris of Philadelphia, even Washington could not have saved the country.

Another source of weakness was the intense dislike and jealousy with which the militia of the different states regarded each other. Their alliance against the common enemy had hitherto done little more toward awakening a cordial sympathy between the states than the alliance of Athenians with Lacedæmonians against the Great King accomplished toward ensuring peace and good-will throughout the Hellenic world. Politically the men of Virginia had thus far acted in remarkable harmony with the men of New England, but socially there was but little fellowship between them. In those days of slow travel the plantations of Virginia were much more remote from Boston than they now are from London, and the generalizations which the one people used to make about the other were, if possible, even more crude than those which Englishmen and Americans are apt to make about each other at the present day. To the stately elegance of the Virginian country mansion it seemed right to sneer at New England merchants and farmers as "shopkeepers" and "peasants," while many people in Boston regarded Virginian planters as mere Squire Westerns. Between the eastern and the middle states, too, there was much ill-will, because of theological differences and boundary disputes. The Puritan of New Hampshire had not yet made up his quarrel with the Churchman of New York concerning the ownership of the Green Mountains; and the wrath of the Pennsylvania Quaker waxed hot against the Puritan of Connecticut who dared claim jurisdiction over the valley of Wyoming. We shall find such animosities bearing bitter fruit in personal squabbles among soldiers and officers, as well as in removals and appointments of officers for reasons which had nothing to do with their military competence. Even in the highest ranks of the army and in Congress these local prejudices played their part and did no end of mischief.

From the outset Washington had laboured with Congress to take measures to obviate these alarming difficulties. In the midst of his retreat through the Jerseys he declared that "short enlistments and a mistaken dependence upon militia have been the origin of all our misfortunes," and at the same time he recommended that a certain number of battalions should be raised directly by the United States, comprising volunteers drawn indiscriminately from the several states. These measures were adopted by Congress, and at the same time Washington was clothed with almost dictatorial powers. It was decided that the army of state troops should be increased to 66,000 men, divided into eighty-eight battalions, of which Massachusetts and Virginia were each to contribute fifteen, "Pennsylvania twelve, North Carolina nine, Connecticut eight, South Carolina six, New York and New Jersey four each, New Hampshire and Maryland three each, Rhode Island two, Delaware and Georgia each one." The actual enlistments fell very far short of this number of men, and the proportions assigned by Congress, based upon the population of the several states, were never heeded. The men now enlisted were to serve during the war, and were to receive at the end a hundred acres of land each as bounty. Colonels were to have a bounty of five hundred acres, and inferior officers were to receive an intermediate quantity. Even with these offers it was found hard to persuade men to enlist for the war, so that it was judged best to allow the recruit his choice of serving for three years and going home empty-handed, or staying till the war should end in the hope of getting a new farm for one of his children. All this enlisting was to be done by the several states, which were also to clothe and arm their recruits, but the money for their equipments, as well as for the payment and support of the troops, was to be furnished by Congress. Officers were to be selected by the states, but formally commissioned by Congress. At the same time Washington was authorized to raise sixteen battalions of infantry, containing 12,000 men, three regiments of artillery, 3,000 light cavalry, and a corps of engineers. These forces were to be enlisted under Washington's direction, in the name of the United States, and were to be taken indiscriminately from all parts of the country. Their officers were to be appointed by Washington, who was furthermore empowered to fill all vacancies and remove any officer below the rank of brigadier-general in any department of the army. Washington was also authorized to take whatever private property might anywhere be needed for the army, allowing a fair compensation to the owners; and he was instructed to arrest at his own discretion, and hold for trial by the civil courts, any person who should refuse to take the continental paper money, or otherwise manifest a want of sympathy with the American cause.

These extraordinary powers, which at the darkest moment of the war were conferred upon Washington for a period of six months, occasioned much grumbling, but it does not appear that any specific difficulty ever arose through the way in which they were exercised. It would be as hard, perhaps, to find any strictly legal justification for the creation of a Continental army as it would be to tell just where the central government of the United States was to be found at that time. Strictly speaking, no central government had as yet been formed. No articles of confederation had yet been adopted by the states, and the authority of the Continental Congress had been in nowise defined. It was generally felt, however, that the Congress now sitting had been chosen for the purpose of representing the states in their relations to the British crown. This Congress had been expressly empowered to declare the states independent of Great Britain, and to wage war for the purpose of making good its declaration. And it was accordingly felt that Congress was tacitly authorized to take such measures as were absolutely needful for the maintenance of the struggle. The enlistment of a Continental force was therefore an act done under an implied "war power," something like the power invoked at a later day to justify the edict by which President Lincoln emancipated the slaves. The thoroughly English political genius of the American people teaches them when and how to tolerate such anomalies, and has more than once enabled them safely to cut the Gordian knot which mere logic could not untie if it were to fumble till doomsday. In the second year after Lexington the American commonwealths had already entered upon the path of their "manifest destiny," and were becoming united into one political body faster than the people could distinctly realize.

CHAPTER VI.

SECOND BLOW AT THE CENTRE.

EVER since the failure of the American invasion of Canada, it had been the intention of Sir Guy Carleton, in accordance with the wishes of the ministry, to invade New York by way of Lake Champlain, and to secure the Mohawk valley and the upper waters of the Hudson. The summer of 1776 had been employed by Carleton in getting together a fleet with which to obtain control of the lake. It was an arduous task. Three large vessels were sent over from England, and proceeded up the St. Lawrence as far as the rapids, where they were taken to pieces, carried overland to St. John's, and there put together again. Twenty gunboats and more than two hundred flat-bottomed transports were built at Montreal, and manned with 700 picked seamen and gunners; and upon this flotilla Carleton embarked his army of 12,000 men.

Carleton invades New York.

To oppose the threatened invasion, Benedict Arnold had been working all the summer with desperate energy. In June the materials for his navy were growing in the forests of Vermont, while his carpenters with their tools, his sailmakers with their canvas, and his gunners with their guns had mostly to be brought from the coast towns of

Connecticut and Massachusetts. By the end of September he had built a little fleet of three schooners, two sloops, three galleys, and eight gondolas, and fitted it out with seventy guns and such seamen and gunners as he could get together. With this flotilla he could not hope to prevent the advance of such an overwhelming force as that of the enemy. The most he could do would be to worry and delay it, besides raising the spirits of the people by the example of an obstinate and furious resistance. To allow Carleton to reach Ticonderoga without opposition would be disheartening, whereas by delay and vexation he might hope to dampen the enthusiasm of the invader. With this end in view, Arnold proceeded down the lake far to the north of Crown Point, and taking up a strong position between Valcour Island and the western shore, so that both his wings were covered and he could be attacked only in front, he lay in wait for the enemy. James Wilkinson, who twenty years afterward became commander-in-chief of the American army, and survived the second war with England, was then at Ticonderoga, on Gates's staff. Though personally hostile to Arnold, he calls attention in his Memoirs to the remarkable skill exhibited in the disposition of the little fleet at Valcour Island, which was the same in principle as that by which Macdonough won his brilliant victory, not far from the same spot, in 1814.

Arnold's preparations.

On the 11th of October, Sir Guy Carleton's squadron approached, and there ensued the first battle fought between an American and a British fleet. At sundown, after a desperate fight of seven hours' duration, the British withdrew out of range, intending to renew the struggle in the morning. Both fleets had suffered severely, but the Americans were so badly cut up that Carleton expected to force their rear the next day, and capture them. But Arnold, during the hazy night, by a feat scarcely less remarkable than Washington's retreat from Long Island, contrived to slip through the British line with all that was left of his crippled flotilla, and made away for Crown Point with all possible speed. Though he once had to stop to mend leaks, and once to take off the men and guns from two gondolas which were sinking, he nevertheless, by dint of sailing and kedging, got such a start that the enemy did not overtake him until the next day but one, when he was nearing Crown Point. While the rest of the fleet, by Arnold's orders, now crowded sail for their haven, he in his schooner sustained an ugly fight for four hours with the three largest British vessels, one of which mounted eighteen twelve-pounders. His vessel was wofully cut up, and her deck covered with dead and dying men, when, having sufficiently delayed the enemy, he succeeded in running her aground in a small creek, where he set her on fire, and she perished gloriously, with her flag flying till the flames brought it down. Then marching through woodland paths to Crown Point, where his other vessels had now disembarked their men, he brought away his whole force in safety to Ticonderoga. When Carleton appeared before that celebrated fortress,

Battle of Valcour Island, Oct. 11, 1776.

finding it strongly defended, and doubting his ability to reduce it before the setting in of cold weather, he decided to take his army back to Canada, satisfied for the present with having gained control of Lake Champlain. This sudden retreat of Carleton astonished both friend and foe. He was blamed for it by his generals, Burgoyne, Phillips, and Riedesel, as well as by the king; and when we see how easily the fortress was seized by Phillips in the following summer, we can hardly doubt that it was a grave mistake.

Arnold had now won an enviable reputation as the "bravest of the brave." In his terrible march through the wilderness of Maine, in the assault upon Quebec, and in the defence of Lake Champlain, he had shown rare heroism and skill. The whole country rang with his praises, and Washington regarded him as one of the ablest officers in the army. Yet when Congress now proceeded to appoint five new major-generals, they selected Stirling, Mifflin, St. Clair, Stephen, and Lincoln, passing over Arnold, who was the senior brigadier. None of the generals named could for a moment be compared with Arnold for ability, and this strange action of Congress, coming soon after such a brilliant exploit, naturally hurt his feelings and greatly incensed him. Arnold was proud and irascible in temper, but on this occasion he controlled himself manfully, and listened to Washington, who entreated him not to resign. So astonished was Washington at the action of Congress that at first he could not believe it. He thought either that

Congress promotes five junior brigadiers over Arnold, Feb. 19, 1777.

Arnold must really have received a prior appointment, which for some reason had not yet been made public, or else that his name must have been omitted through some unaccountable oversight. It turned out, however, on further inquiry, that state jealousies had been the cause of the mischief. The reason assigned for ignoring Arnold's services was that Connecticut had already two major-generals, and was not in fairness entitled to any more! But beneath this alleged reason there lurked a deeper reason, likewise founded in jealousies between the states. The intrigues which soon after disgraced the northern army and imperilled the safety of the country had already begun to bear bitter fruit. Since the beginning of the war, Major-General Philip Schuyler had been in command of the northern department, with his headquarters at Albany, whence his ancestors had a century before hurled defiance at Frontenac. His family was one of the most distinguished in New York, and an inherited zeal for the public service thrilled in every drop of his blood. No more upright or disinterested man could be found in America, and for bravery and generosity he was like the paladin of some mediæval romance. In spite of these fine qualities, he was bitterly hated by the New England men, who formed a considerable portion of his army. Beside the general stupid dislike which the people of New York and of New England then felt for each other, echoes of which are still sometimes heard nowadays, there was a special reason for the odium which was heaped upon Schuyler. The dispute over

Philip Schuyler.

the possession of Vermont had now raged fiercely for thirteen years, and Schuyler, as a member of the New York legislature, had naturally been zealous in urging the claims of his own state. For this crime the men of New England were never able to forgive him, and he was pursued with vindictive hatred until his career as a general was ruined. His orders were obeyed with sullenness, the worst interpretation was put upon every one of his acts, and evil-minded busybodies were continually pouring into the ears of Congress a stream of tattle, which gradually wore out their trust in him.

The evil was greatly enhanced by the fact that among the generals of the northern army there was one envious creature who was likely to take Schuyler's place in case he should be ousted from it, and who for so desirable an object was ready to do any amount of intriguing. The part sustained by Charles Lee with reference to Washington was to some extent paralleled here by the part sustained toward Schuyler by Horatio Gates. There is indeed no reason for supposing that Gates was capable of such baseness as Lee exhibited in his willingness to play into the hands of the enemy; nor had he the nerve for such prodigious treason as that in which Arnold engaged after his sympathies had become alienated from the American cause. With all his faults, Gates never incurred the odium which belongs to a public traitor. But his nature was thoroughly weak and petty, and he never shrank from falsehood when it seemed to serve his purpose. Unlike Lee,

Horatio Gates.

he was comely in person, mild in disposition, and courteous in manner, except when roused to anger or influenced by spite, when he sometimes became very violent. He never gave evidence either of skill or of bravery; and in taking part in the war his only solicitude seems to have been for his own personal advancement. In the course of his campaigning with the northern army, he seems never once to have been under fire, but he would incur no end of fatigue to get a private talk with a delegate in Congress. Like many others, he took a high position at the beginning of the struggle simply because he was a veteran of the Seven Years' War, having been one of the officers who were brought off in safety from the wreck of Braddock's army by the youthful skill and prowess of Washington. At present, and until after the end of the Saratoga campaign, such reputation as he had was won by appropriating the fame which was earned by his fellow-generals. He was in command at Ticonderoga when Arnold performed his venturesome feat on Lake Champlain, and when Carleton made his blunder in not attacking the stronghold; and all this story Gates told to Congress as the story of an advantage which he had somehow gained over Carleton, at the same time anxiously inquiring if Congress regarded him, in his remote position at Ticonderoga, as subject to the orders of Schuyler at Albany. Finding that he was thus regarded as subordinate, he became restive, and seized the earliest opportunity of making a visit to Congress. The retreat of Carleton enabled Schuyler to send seven regiments to the relief of Washington in

New Jersey, and we have already seen how Gates, on arriving with this reinforcement, declined to assist personally in the Trenton campaign, and took the occasion to follow Congress in its retreat to Baltimore.

The winter seems to have been spent in intrigue. Knowing the chief source of Schuyler's unpopularity, Gates made it a point to declare, as often and as loudly as possible, his belief that the state of New York had no title to the Green Mountain country. In this way he won golden opinions from the people of New England, and rose high in the good graces of such members of Congress as Samuel Adams, whose noble nature was slow to perceive his meanness and duplicity. The failure of the invasion of Canada had caused much chagrin in Congress, and it was sought to throw the whole blame of it upon Schuyler for having, as it was alleged, inadequately supported Montgomery and Arnold. The unjust charge served to arouse a prejudice in many minds, and during the winter some irritated letters passed between Schuyler and Congress, until late in March, 1777, he obtained permission to visit Philadelphia and vindicate himself. On the 22d of May, after a thorough investigation, Schuyler's conduct received the full approval of Congress, and he was confirmed in his command of the northern department, which was expressly defined as including Lakes George and Champlain, as well as the valleys of the Hudson and the Mohawk.

Gates intrigues against Schuyler.

The sensitive soul of Gates now took fresh offence. He had been sent back in March to his post at Ticonderoga, just as Schuyler was starting for Philadelphia, and he flattered himself with the hope that he would soon be chosen to supersede his gallant commander. Accordingly when he found that Schuyler had been reinstated in all his old command and honours, he flew into a rage, refused to serve in a subordinate capacity, wrote an impudent letter to Washington, and at last got permission to visit Congress again, while General St. Clair was appointed in his stead to the command of the great northern fortress. On the 19th of June, Gates obtained a hearing before Congress, and behaved with such unseemly violence that after being repeatedly called to order, he was turned out of the room, amid a scene of angry confusion. Such conduct should naturally have ruined his cause, but he had made so many powerful friends that by dint of more or less apologetic talk the offence was condoned.

Gates visits Congress.

Throughout these bickerings Arnold had been the steadfast friend of Schuyler; and although his brilliant exploits had won general admiration he did not fail to catch some of the odium so plentifully bestowed upon the New York commander. In the chaos of disappointment and wrath which ensued upon the disastrous retreat from Canada in 1776, when everybody was eager to punish somebody else for the ill fortune which was solely due to the superior resources of the enemy, Arnold came in for his share of blame. No one could find any fault with his military conduct, but charges were brought against him on the ground of some exactions of private property at

Charges against Arnold.

Montreal which had been made for the support of the army. A thorough investigation of the case demonstrated Arnold's entire uprightness in the matter, and the verdict of Congress, which declared the charges to be "cruel and unjust," was heartily indorsed by Washington. Nevertheless, in the manifold complications of feeling which surrounded the Schuyler trouble, these unjust charges succeeded in arousing a prejudice which may have had something to do with the slight cast upon Arnold in the appointment of the new major-generals. In the whole course of American history there are few sadder chapters than this. Among the scandals of this eventful winter we can trace the beginnings of the melancholy chain of events which by and by resulted in making the once heroic name of Benedict Arnold a name of opprobrium throughout the world. We already begin to see, too, originating in Lee's intrigues of the preceding autumn, and nourished by the troubles growing out of the Vermont quarrel and the ambitious schemes of Gates, the earliest germs of that faction which erelong was to seek to compass the overthrow of Washington himself.

For the present the injustice suffered by Arnold had not wrought its darksome change in him. A long and complicated series of influences was required to produce that result. To the earnest appeal of Washington that he should not resign he responded cordially, declaring that no personal considerations should induce him to stay at home while the interests of his country were at stake. He would zealously serve even under his juniors, who had lately been raised above him, so long as the common welfare was in danger. An opportunity for active service soon presented itself. Among the preparations for the coming summer campaign, Sir William Howe thought it desirable to cripple the Americans by seizing a large quantity of military stores which had been accumulated at Danbury in Connecticut. An expedition was sent out, very much like that which at Lexington and Concord had ushered in the war, and it met with a similar reception. A force of 2,000 men, led by the royal governor, Tryon, of North Carolina fame, landed at Fairfield, and marched to Danbury, where they destroyed the stores and burned a large part of the town. The militia turned out, as on the day of Lexington, led by General Wooster, who was slain in the first skirmish. By this time Arnold, who happened to be visiting his children in New Haven, had heard of the affair, and came upon the scene with 600 men. At Ridgefield a desperate fight ensued, in which Arnold had two horses killed under him. The British were defeated. By the time they reached their ships, 200 of their number had been killed or wounded, and, with the yeomanry swarming on every side, they narrowly escaped capture. For his share in this action Arnold was now made a major-general, and was presented by Congress with a fine horse; but nothing was done towards restoring him to his relative rank, nor was any explanation vouchsafed. Washington offered him the command of the Hudson at Peekskill, which was

Tryon's expedition against Danbury.

Arnold defeats Tryon at Ridgefield, April 27, 1777.

liable to prove one of the important points in the ensuing campaign; but Arnold for the moment declined to take any such position until he should have conferred with Congress, and fathomed the nature of the difficulties by which he had been beset; and so the command of this important position was given to the veteran Putnam.

The time for the summer campaign was now at hand. The first year of the independence of the United States was nearly completed, and up to this time the British had nothing to show for their work except the capture of the city of New York and the occupation of Newport. The army of Washington, which six months ago they had regarded as conquered and dispersed, still balked and threatened them from its inexpugnable position on the heights of Morristown. It was high time that something more solid should be accomplished, for every month of adverse possession added fresh weight to the American cause, and increased the probability that France would interfere.

A decisive blow was accordingly about to be struck. After careful study by Lord George Germain, and much consultation with General Burgoyne, who had returned to England for the winter, it was decided to adhere to the plan of the preceding year, with slight modifications. The great object was to secure firm possession of the entire valley of the Hudson, together with that of the Mohawk. It must be borne in mind that at this time the inhabited part of the state of New York consisted almost entirely of the Mohawk and Hudson valleys. All the rest was unbroken wilderness, save for an occasional fortified trading-post. With a total population of about 170,000, New York ranked seventh among the thirteen states; just after Maryland and Connecticut, just before South Carolina. At the same time, the geographical position of New York, whether from a commercial or from a military point of view, was as commanding then as it has ever been. It was thought that so small a population, among which there were known to be many Tories, might easily be conquered and the country firmly held. The people of New Jersey and Pennsylvania were regarded as lukewarm supporters of the Declaration of Independence, and it was supposed that the conquest of New York might soon be followed by the subjection of these two provinces. With the British power thus thrust, like a vast wedge, through the centre of the confederacy, it would be impossible for New England to coöperate with the southern states, and it was hoped that the union of the colonies against the Crown would thus be effectually broken.

The military centre of the United States was the state of New York.

With this object of conquering New York, we have seen Carleton, in 1776, approaching through Lake Champlain, while Howe was wresting Manhattan Island from Washington. But the plan was imperfectly conceived, and the coöperation was feeble. How feeble it was is well shown by the fact that Carleton's ill-judged retreat from Crown Point enabled Schuyler to send reinforcements to Washington in time to take part in the great strokes at Trenton and Princeton. Something, however, had been accomplished. In spite of Arnold's desperate resistance and Washington's consummate skill, the enemy had gained a hold upon both the northern and the southern ends of the long line. But this obstinate resistance served to some extent to awaken the enemy to the arduous character of the problem. The plan was more carefully studied, and it was intended that this time the coöperation should be more effectual. In order to take possession of the whole state by one grand system of operations, it was decided that the invasion should be conducted by three distinct armies operating upon converging lines. A strong force from Canada was to take Ticonderoga, and proceed down the line of the Hudson to Albany. This force was now to be commanded by General Burgoyne, while his superior officer, General Carleton, remained at Quebec. A second and much smaller force, under Colonel St. Leger, was to go up the St. Lawrence to Lake Ontario, land at Oswego, and, with the aid of Sir John Johnson and the Indians, reduce Fort Stanwix; after which he was to come down the Mohawk valley and unite his forces with those of Burgoyne. At the same time, Sir William Howe was to ascend the Hudson with the main army, force the passes of the Highlands at Peekskill, and effect a junction with Burgoyne at Albany. The junction of the three armies was expected to complete the conquest of New York, and to insure the overthrow of American independence.

A second blow to be struck at the centre. The plan of campaign.

Such was the plan of campaign prepared by the

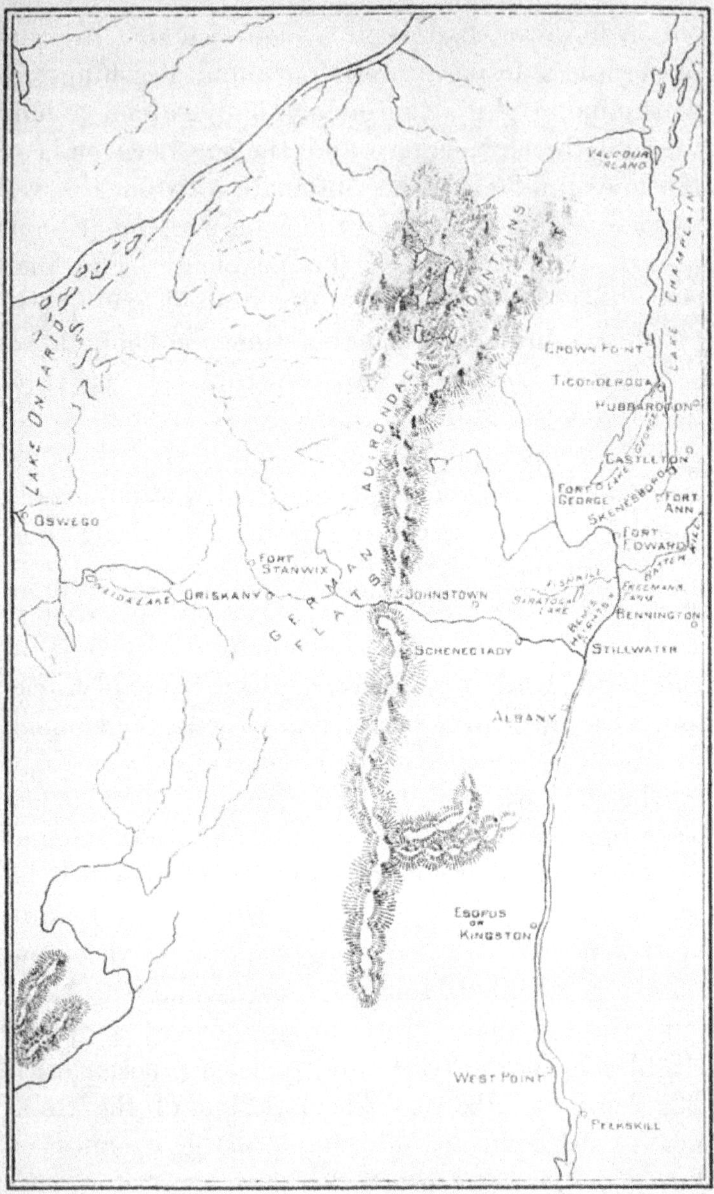

BURGOYNE'S CAMPAIGN

JULY — OCTOBER, 1777

ministry. There can be no doubt that it was carefully studied, or that, if successful, it would have proved very disastrous to the Americans. There is room for very grave doubt, however, as to whether it was the most judicious plan to adopt. The method of invading any country by distinct forces operating upon converging lines is open to the objection that either force is liable to be separately overwhelmed without the possibility of reinforcement from the other. Such a plan is prudent only when the invaded country has good roads, and when the invaders have a great superiority in force, as was the case when the allied armies advanced upon Paris in 1814. In northern and central New York, in 1777, the conditions were very unfavourable to such a plan. The distances to be traversed were long, and the roads were few and bad. Except in the immediate neighbourhood of Albany and Saratoga, the country was covered with the primeval forest, through which only the trapper and the savage could make their way with speed. The Americans, too, had the great advantage of operating upon interior lines. It was difficult for Burgoyne at Fort Edward, St. Leger before Fort Stanwix, and Howe in the city of New York to communicate with each other at all; it was impossible for them to do so promptly; whereas nothing could be easier than for Washington at Morristown to reach Putnam at Peekskill, or for Putnam to forward troops to Schuyler at Albany, or for Schuyler to send out a force to raise the siege of Fort Stanwix. In view of these considerations, it seems probable that

The plan was unsound.

Lord George Germain would have acted more wisely if he had sent Burgoyne with his army directly by sea to reinforce Sir William Howe. The army thus united, and numbering more than 30,000 men, would have been really formidable. If they had undertaken to go up the river to Albany, it would have been hard to prevent them. If their united presence at Albany was the great object of the campaign, there was no advantage in sending one commander to reach it by a difficult and dangerous overland march. The Hudson is navigable by large vessels all the way to Albany, and by advancing in this way the army might have preserved its connections; and whatever disaster might have befallen, it would have been difficult for the Americans to surround and capture so large a force. Once arrived at Albany, the expedition of St. Leger might have set out from that point as a matter of subsequent detail, and would have had a base within easy distance upon which to fall back in case of defeat.

It does not appear, therefore, that there were any advantages to be gained by Burgoyne's advance from the north which can be regarded as commensurate with the risk which he incurred. To have transferred the northern army from the St. Lawrence to the Hudson by sea would have been far easier and safer than to send it through a hundred miles of wilderness in northern New York; and whatever it could have effected in the interior of the state could have been done as well in the former case as in the latter. But these considerations do not seem to have occurred to Lord

George Germain. In the wars with the French, the invading armies from Canada had always come by way of Lake Champlain, so that this route was accepted without question, as if consecrated by long usage. Through a similar association of ideas an exaggerated importance was attached to the possession of Ticonderoga. The risks of the enterprise, moreover, were greatly underestimated. In imagining that the routes of Burgoyne and St. Leger would lie through a friendly country, the ministry fatally misconceived the whole case. There was, indeed, a powerful Tory party in the country, just as in the days of Robert Bruce there was an English party in Scotland, just as in the days of Miltiades there was a Persian party in Attika. But no one has ever doubted that the victors at Marathon and at Bannockburn went forth with a hearty godspeed from their fellow-countrymen; and the obstinate resistance encountered by St. Leger, within a short distance of Johnson's Tory stronghold, is an eloquent commentary upon the error of the ministry in the estimate of the actual significance of the loyalist element on the New York frontier.

Germaine's fatal error.

It thus appears that in the plan of a triple invasion upon converging lines the ministry were dealing with too many unknown quantities. They were running a prodigious risk for the sake of an advantage which in itself was extremely open to question; for should it turn out that the strength of the Tory party was not sufficiently great to make the junction of the three armies at Albany at once equivalent to the

Too many unknown quantities.

complete conquest of the state, then the end for which the campaign was undertaken could not be secured without supplementary campaigns. Neither a successful march up and down the Hudson river nor the erection of a chain of British fortresses on that river could effectually cut off the southern communications of New England, unless all military resistance were finally crushed in the state of New York. The surest course for the British, therefore, would have been to concentrate all their available force at the mouth of the Hudson, and continue to make the destruction of Washington's army the chief object of their exertions. In view of the subtle genius which he had shown during the last campaign, that would have been an arduous task; but, as events showed, they had to deal with his genius all the same on the plan which they adopted, and at a great disadvantage.

Danger from New England ignored.

Another point which the ministry overlooked was the effect of Burgoyne's advance upon the people of New England. They could reasonably count upon alarming the yeomanry of New Hampshire and Massachusetts by a bold stroke upon the Hudson, but they failed to see that this alarm would naturally bring about a rising that would be very dangerous to the British cause. Difficult as it was at that time to keep the Continental army properly recruited, it was not at all difficult to arouse the yeomanry in the presence of an immediate danger. In the western parts of New England there were scarcely any Tories to complicate the matter; and the flank movement by the New England militia became one of the most formidable features in the case.

But whatever may be thought of the merits of Lord George's plan, there can be no doubt that its success was absolutely dependent upon the harmonious coöperation of all the forces involved in it. The ascent of the Hudson by Sir William Howe, with the main army, was as essential a part of the scheme as the descent of Burgoyne from the north; and as the two commanders could not easily communicate with each other, it was necessary that both should be strictly bound by their instructions. At this point a fatal blunder was made. Burgoyne was expressly directed to follow the prescribed line down the Hudson, whatever might happen, until he should effect his junction with the main army. On the other hand, no such unconditional orders were received by Howe. He understood the plan of campaign, and knew that he was expected to ascend the river in force; but he was left with the usual discretionary power, and we shall presently see what an imprudent use he made of it. The reasons for this inconsistency on the part of the ministry were for a long time unintelligible; but a memorandum of Lord Shelburne, lately brought to light by Lord Edmund Fitzmaurice, has solved the mystery. It seems that a dispatch, containing positive and explicit orders for Howe to ascend the Hudson, was duly drafted, and, with many other papers, awaited the minister's signature. Lord George Germain, being on his way to the country, called at his office to sign the dispatches; but when he came to the letter addressed to General Howe, he found it had not been "fair copied."

The dispatch that was never sent.

Lord George, like the old gentleman who killed himself in defence of the great principle that crumpets are wholesome, never would be put out of his way by anything. Unwilling to lose his holiday he hurried off to the green meadows of Kent, intending to sign the letter on his return. But when he came back the matter had slipped from his mind. The document on which hung the fortunes of an army, and perhaps of a nation, got thrust unsigned into a pigeon-hole, where it was duly discovered some time after the disaster at Saratoga had become part of history.

Happy in his ignorance of the risks he was assuming, Burgoyne took the field about the 1st of June, with an army of 7,902 men, of whom 4,135 were British regulars. His German troops from Brunswick, 3,116 in number, were commanded by Baron Riedesel, an able general, whose accomplished wife has left us such a picturesque and charming description of the scenes of this adventurous campaign. Of Canadian militia there were 148, and of Indians 503. The regular troops, both German and English, were superbly trained and equipped, and their officers were selected with especial care. Generals Phillips and Fraser were regarded as among the best officers in the British service. On the second anniversary of Bunker Hill this army began crossing the lake to Crown Point; and on the 1st of July it appeared before Ticonderoga, where St. Clair was posted with a garrison of 3,000 men. Since its capture by Allen, the fortress had been

Burgoyne advances upon Ticonderoga.

carefully strengthened, until it was now believed to be impregnable. But while no end of time and expense had been devoted to the fortifications, a neighbouring point which commands the whole position had been strangely neglected. A little less than a mile south of Ticonderoga, the narrow mountain ridge between the two lakes ends abruptly in a bold crag, which rises 600 feet sheer over the blue water. Practised eyes in the American fort had already seen that a hostile battery planted on this eminence would render their stronghold untenable; but it was not believed that siege-guns could be dragged up the steep ascent, and so, in spite of due warning, the crag had not been secured when the British army arrived. General Phillips at once saw the value of the position, and, approaching it by a defile that was screened from the view of the fort, worked night and day in breaking out a pathway and dragging up cannon. "Where a goat can go, a man may go; and where a man can go, he can haul up a gun," argued the gallant general. Great was the astonishment of the garrison when, on the morning of July 5th, they saw red coats swarming on the top of the hill, which the British, rejoicing in their exploit, now named Mount Defiance. There were not only red coats there, but brass cannon, which by the next day would be ready for work. Ticonderoga had become a trap, from which the garrison could not escape too quickly. A council of war was held, and under cover of night St. Clair took his little army across the lake and retreated upon Castleton

Phillips seizes Mount Defiance.

St. Clair abandons Ticonderoga, July 5, 1777.

in the Green Mountains. Such guns and stores as could be saved, with the women and wounded men, were embarked in 200 boats, and sent, under a strong escort, to the head of the lake, whence they continued their retreat to Fort Edward on the Hudson. About three o'clock in the morning a house accidentally took fire, and in the glare of the flames the British sentinels caught a glimpse of the American rear-guard just as it was vanishing in the sombre depths of the forest. Alarm guns were fired, and in less than an hour the British flag was hoisted over the empty fortress, while General Fraser, with 900 men, had started in hot pursuit of the retreating Americans. Riedesel was soon sent to support him, while Burgoyne, leaving nearly 1,000 men to garrison the fort, started up the lake with the main body of the army.

Battle of Hubbardton, July 7.

On the morning of the 7th, General Fraser overtook the American rear-guard of 1,000 men, under Colonels Warner and Francis, at the village of Hubbardton, about six miles behind the main army. A fierce fight ensued, in which Fraser was worsted, and had begun to fall back, with the loss of one fifth of his men, when Riedesel came up with his Germans, and the Americans were put to flight, leaving one third of their number killed or wounded. This obstinate resistance at Hubbardton served to check the pursuit, and five days later St. Clair succeeded, without further loss, in reaching Fort Edward, where he joined the main army under Schuyler.

Up to this moment, considering the amount of

work done and the extent of country traversed, the loss of the British had been very small. They began to speak contemptuously of their antagonists, and the officers amused themselves by laying wagers as to the precise number of days it would take them to reach Albany. In commenting on the failure to occupy Mount Defiance, Burgoyne made a general statement on the strength of a single instance, — which is the besetting sin of human reasoning. "It convinces me," said he, "that the Americans have no men of military science." Yet General Howe at Boston, in neglecting to occupy Dorchester Heights, had made just the same blunder, and with less excuse; for no one had ever doubted that batteries might be placed there by somebody.

One swallow does not make a summer.

In England the fall of Ticonderoga was greeted with exultation, as the death-blow to the American cause. Horace Walpole tells how the king rushed into the queen's apartment, clapping his hands and shouting, "I have beat them! I have beat all the Americans!" People began to discuss the best method of reëstablishing the royal governments in the "colonies." In America there was general consternation. St. Clair was greeted with a storm of abuse. John Adams, then president of the Board of War, wrote, in the first white heat of indignation, "We shall never be able to defend a post till we shoot a general!" Schuyler, too, as commander of the department, was ignorantly and wildly blamed, and his political enemies seized upon the occasion to circulate fresh stories to his

The king's glee.

Wrath of John Adams.

discredit. A court-martial in the following year vindicated St. Clair's prudence in giving up an untenable position and saving his army from capture. The verdict was just, but there is no doubt that the failure to fortify Mount Defiance was a grave error of judgment, for which the historian may fairly apportion the blame between St. Clair and Gates. It was Gates who had been in command of Ticonderoga in the autumn of 1776, when an attack by Carleton was expected, and his attention had been called to this weak point by Colonel Trumbull, whom he laughed to scorn. Gates had again been in command from March to June. St. Clair had taken command about three weeks before Burgoyne's approach; he had seriously considered the question of fortifying Mount Defiance, but had not been sufficiently prompt. In no case could any blame attach to Schuyler. Gates was more at fault than any one else, but he did not happen to be at hand when the catastrophe occurred, and accordingly people did not associate him with it. On the contrary, amid the general wrath, the loss of the northern citadel was alleged as a reason for superseding Schuyler by Gates; for if he had been there, it was thought that the disaster would have been prevented.

Gates chiefly to blame.

The irony of events, however, alike ignoring American consternation and British glee, showed that the capture of Ticonderoga was not to help the invaders in the least. On the contrary, it straightway became a burden, for it detained an eighth part of Burgoyne's force in garrison at a time when he could ill spare it. Indeed, alarming as his swift advance had seemed at first, Burgoyne's serious difficulties were now just beginning, and the harder he laboured to surmount them the more completely did he work himself into a position from which it was impossible either to advance or to recede. On the 10th of July his whole army had reached Skenesborough (now Whitehall), at the head of Lake Champlain. From this point to Fort Edward, where the American army was encamped, the distance was twenty miles as the crow flies; but Schuyler had been industriously at work with those humble weapons the axe and the crowbar, which in warfare sometimes prove mightier than the sword. The roads, bad enough at their best, were obstructed every few yards by huge trunks of fallen trees, that lay with their boughs interwoven. Wherever the little streams could serve as aids to the march, they were choked up with stumps and stones; wherever they served as obstacles which needed to be crossed, the bridges were broken down. The country was such an intricate labyrinth of creeks and swamps that more than forty bridges had to be rebuilt in the course of the march. Under these circumstances, Burgoyne's advance must be regarded as a marvel of celerity. He accomplished a mile a day, and reached Fort Edward on the 30th of July.

Burgoyne's difficulties begin.

In the mean time Schuyler had crossed the Hudson, and slowly fallen back to Stillwater, about thirty miles above Albany. For this retrograde movement fresh blame was visited upon him by the general public,

Schuyler wisely evacuates Fort Edward.

which at all times is apt to suppose that a war should mainly consist of bloody battles, and which can seldom be made to understand the strategic value of a retreat. The facts of the case were also misunderstood. Fort Edward was supposed to be an impregnable stronghold, whereas it was really commanded by highlands. The Marquis de Chastellux, who visited it somewhat later, declared that it could be taken at any time by 500 men with four siege-guns. Now for fighting purposes an open field is much better than an untenable fortress. If Schuyler had stayed in Fort Edward, he would probably have been forced to surrender; and his wisdom in retreating is further shown by the fact that every moment of delay counted in his favour. The militia of New York and New England were already beating to arms. Some of those yeomen who were with the army were allowed to go home for the harvest; but the loss was more than made good by the numerous levies which, at Schuyler's suggestion and by Washington's orders, were collecting under General Lincoln in Vermont, for the purpose of threatening Burgoyne in the rear. The people whose territory was invaded grew daily more troublesome to the enemy. Burgoyne had supposed that it would be necessary only to show himself at the head of an army, when the people would rush by hundreds to offer support or seek protection. He now found that the people withdrew from his line of advance, driving their cattle before them, and seeking shelter, when possible, within the lines of the American army. In his reliance upon the aid of New

Enemies gathering in Burgoyne's rear.

York loyalists, he was utterly disappointed; very few Tories joined him, and these could offer neither sound advice nor personal influence wherewith to help him. When the yeomanry collected by hundreds, it was only to vex him and retard his progress.

Even had the loyalist feeling on the Vermont frontier of New York been far stronger than it really was, Burgoyne had done much to alienate or stifle it by his ill-advised employment of Indian auxiliaries. For this blunder the responsibility rests mainly with Lord North and Lord George Germaine. Burgoyne had little choice in the matter except to carry out his instructions. Being a humane man, and sharing, perhaps, in that view of the "noble savage" which was fashionable in Europe in the eighteenth century, he fancied he could prevail upon his tawny allies to forego their cherished pastime of murdering and scalping. When, at the beginning of the campaign, he was joined by a party of Wyandots and Ottawas, under command of that same redoubtable Charles de Langlade who, twenty-two years before, had achieved the ruin of Braddock, he explained his policy to them in an elaborate speech, full of such sentimental phrases as the Indian mind was supposed to delight in. The slaughter of aged men, of women and children and unresisting prisoners, was absolutely prohibited; and "on no account, or pretense, or subtlety, or prevarication," were scalps to be taken from wounded or dying men. An order more likely to prove efficient was

Use of Indian auxiliaries.

Burgoyne's address to the chiefs.

one which provided a reward for every savage who should bring his prisoners to camp in safety. To these injunctions, which must have inspired them with pitying contempt, the chiefs laconically replied that they had "sharpened their hatchets upon their affections," and were ready to follow their "great white father."

The employment of savage auxiliaries was indignantly denounced by the opposition in Parliament, and when the news of this speech of Burgoyne's reached England it was angrily ridiculed by Burke, who took a sounder view of the natural instincts of the red man. "Suppose," said Burke, "that there was a riot on Tower Hill. What would the keeper of his majesty's lions do? Would he not fling open the dens of the wild beasts, and then address them thus? 'My gentle lions, my humane bears, my tender-hearted hyenas, go forth! But I exhort you, as you are Christians and members of civilized society, to take care not to hurt any man, woman, or child!'" The House of Commons was convulsed over this grotesque picture; and Lord North, to whom it seemed irresistibly funny to hear an absent man thus denounced for measures which he himself had originated, sat choking with laughter, while tears rolled down his great fat cheeks.

It is ridiculed by Burke.

It soon turned out, however, to be no laughing matter. The cruelties inflicted indiscriminately upon patriots and loyalists soon served to madden the yeomanry, and array against the invaders whatever wavering sentiment had hitherto remained in the country. One sad incident in particular has been treasured up in the memory of the people, and celebrated in song and story. Jenny McCrea, the beautiful daughter of a Scotch clergyman of Paulus Hook, was at Fort Edward, visiting her friend Mrs. McNeil, who was a loyalist and a cousin of General Fraser. On the morning of July 27th, a marauding party of Indians burst into the house, and carried away the two ladies. They were soon pursued by some American soldiers, who exchanged a few shots with them. In the confusion which ensued the party was scattered, and Mrs. McNeil was taken alone into the camp of the approaching British army. Next day a savage of gigantic stature, a famous sachem, known as the Wyandot Panther, came into the camp with a scalp which Mrs. McNeil at once recognized as Jenny's, from the silky black tresses, more than a yard in length. A search was made, and the body of the poor girl was found hard by a spring in the forest, pierced with three bullet wounds. How she came to her cruel death was never known. The Panther plausibly declared that she had been accidentally shot during the scuffle with the soldiers, but his veracity was open to question, and the few facts that were known left ample room for conjecture. The popular imagination soon framed its story with a romantic completeness that thrust aside even these few facts. Miss McCrea was betrothed to David Jones, a loyalist who was serving as lieutenant in Burgoyne's army. In the legend which immediately sprang up, Mr. Jones was said to have sent a party of Indians, with a letter to his betrothed,

The story of Jane McCrea.

entreating her to come to him within the British lines that they might be married. For bringing her to him in safety the Indians were to receive a barrel of rum. When she had entrusted herself to their care, and the party had proceeded as far as the spring, where the savages stopped to drink, a dispute arose as to who was to have the custody of the barrel of rum, and many high words ensued, until one of the party settled the question offhand by slaying the lady with his tomahawk. It would be hard to find a more interesting example of the mushroom-like growth and obstinate vitality of a romantic legend. The story seems to have had nothing in common with the observed facts, except the existence of the two lovers and the Indians and a spring in the forest.[1] Yet it took possession of the popular mind almost immediately after the event, and it has ever since been repeated, with endless variations in detail, by American historians. Mr. Jones himself — who lived, a broken-hearted

[1] I leave this as I wrote it in June, 1883. Since then another version of the facts has been suggested by W. L. Stone in Appleton's *Cyclopædia of American Biography*. In this version, Mr. Jones sends a party of Indians under the half-breed Duluth to escort Miss McCrea to the camp, where they are to be married by Mr. Brudenell, the chaplain. It is to be quite a fine little wedding, and the Baroness Riedesel and Lady Harriet Ackland are to be among the spectators. Before Duluth reaches Mrs. McNeil's house, the Wyandot Panther (here known by the name of a different beast, Le Loup) with his party attacks the house and carries off the two ladies. The Panther's party meets Duluth's near the spring. Duluth insists upon taking Jenny with him, and high words ensue between him and the Panther, until the latter, in a towering rage, draws his pistol and shoots the girl. This version, if correct, goes some way toward reconciling the legend with the observed facts.

man, for half a century after the tragedy — was never weary of pointing out its falsehood and absurdity; but all his testimony, together with that of Mrs. McNeil and other witnesses, to the facts that really happened was powerless to shake the hold upon the popular fancy which the legend had instantly gained. Such an instance, occurring in a community of shrewd and well-educated people, affords a suggestive commentary upon the origin and growth of popular tales in earlier and more ignorant ages.

But in whatever way poor Jenny may have come to her death, there can be no doubt as to the mischief which it swiftly wrought for the invading army. In the first place, it led to the desertion of all the savage allies. Burgoyne was a man of quick and tender sympathy, and the fate of this sweet young lady shocked him as it shocked the American people. He would have had the Panther promptly hanged, but that his guilt was not clearly proved, and many of the officers argued that the execution of a famous and popular sachem would enrage all the other Indians, and might endanger the lives of many of the soldiers. The Panther's life was accordingly spared, but Burgoyne made it a rule that henceforth no party of Indians should be allowed to go marauding save under the lead of some British officer, who might watch and restrain them. When this rule was put in force, the tawny savages grunted and growled for two or three days, and then, with hoarse yells and hoots, all the five hundred broke loose from the camp, and scampered

The Indians desert Burgoyne.

off to the Adirondack wilderness. From a military point of view, the loss was small, save in so far as it deprived the army of valuable scouts and guides. But the thirst for vengeance which was aroused among the yeomanry of northern New York, of Vermont, and of western Massachusetts was a much more serious matter. The lamentable story was told at every village fireside, and no detail of pathos or of horror was forgotten. The name of Jenny McCrea became a watchword, and a fortnight had not passed before General Lincoln had gathered on the British flank an army of stout and resolute farmers, inflamed with such wrath as had not filled their bosoms since the day when all New England had rushed to besiege the enemy in Boston.

Such a force of untrained yeomanry is of little use in prolonged warfare, but on important occasions it is sometimes capable of dealing heavy blows. We have seen what it could do on the memorable day of Lexington. It was now about to strike, at a critical moment, with still more deadly effect. Burgoyne's advance, laborious as it had been for the last three weeks, was now stopped for want of horses to drag the cannon and carry the provision bags; and the army, moreover, was already suffering from hunger. The little village of Bennington, at the foot of the Green Mountains, had been selected by the New England militia as a centre of supplies. Many hundred horses had been collected there, with ample stores of food and ammunition. To capture this village would give Burgoyne the warlike material he wanted, while at

Importance of Bennington; Burgoyne sends a German force against it.

the same time it would paralyze the movements of Lincoln, and perhaps dispel the ominous cloud that was gathering over the rear of the British army. Accordingly, on the 13th of August, a strong detachment of 500 of Riedesel's men, with 100 newly arrived Indians and a couple of cannon, was sent out to seize the stores at Bennington. Lieutenant-Colonel Baum commanded the expedition, and he was accompanied by Major Skene, an American loyalist, who assured Burgoyne on his honour that the Green Mountains were swarming with devoted subjects of King George, who would flock by hundreds to his standard as soon as it should be set up among them. That these loyal recruits might be organized as quickly as possible, Burgoyne sent along with the expedition a skeleton regiment of loyalists, all duly officered, into the ranks of which they might be mustered without delay. The loyal recruits, however, turned out to be the phantom of a distempered imagination: not one of them appeared in the flesh. On the contrary, the demeanour of the people was so threatening that Baum became convinced that hard work was before him, and next day he sent back for reinforcements. Lieutenant-Colonel Breymann was accordingly sent to support him, with another body of 500 Germans and two field-pieces.

Meanwhile Colonel Stark was preparing a warm reception for the invaders. We have already seen John Stark, a gallant veteran of the Seven Years' War, serving with distinction at Bunker Hill and at Trenton and Princeton. He was considered one of the ablest officers in the army; but he had lately gone home in disgust, for, like Arnold, he had been passed over by Congress in the list of promotions. Tired of sulking in his tent, no sooner did this rustic Achilles hear of the invaders' presence in New England than he forthwith sprang to arms, and in the twinkling of an eye 800 stout yeomen were marching under his orders. He refused to take instructions from any superior officer, but declared that he was acting under the sovereignty of New Hampshire alone, and would proceed upon his own responsibility in defending the common cause. At the same time he sent word to General Lincoln, at Manchester in the Green Mountains, asking him to lend him the services of Colonel Seth Warner, with the gallant regiment which had checked the advance of Fraser at Hubbardton. Lincoln sent the reinforcement without delay, and after marching all night in a drenching rain, the men reached Bennington in the morning, wet to the skin. Telling them to follow him as soon as they should have dried and rested themselves, Stark pushed on, with his main body, and found the enemy about six miles distant. On meeting this large force, Baum hastily took up a strong position on some rising ground behind a small stream, everywhere fordable, known as the Walloomsac river. All day long the rain fell in torrents, and while the Germans began to throw up intrenchments, Stark laid his plans for storming their position on the morrow. During the night a company of Berkshire militia arrived, and with them the excellent Mr. Allen, the warlike parson of Pittsfield, who went up to

Stark prepares to receive the Germans.

Stark and said, "Colonel, our Berkshire people have been often called out to no purpose, and if you don't let them fight now they will never turn out again." "Well," said Stark, "would you have us turn out now, while it is pitch dark and raining buckets?" "No, not just this minute," replied the minister. "Then," said the doughty Stark, "as soon as the Lord shall once more send us sunshine, if I don't give you fighting enough, I'll never ask you to come out again!"

Next morning the sun rose bright and clear, and a steam came up from the sodden fields. It was a true dog-day, sultry and scorching. The forenoon was taken up in preparing the attack, while Baum waited in his strong position. The New Englanders outnumbered the Germans two to one, but they were a militia, unfurnished with bayonets or cannon, while Baum's soldiers were all regulars, picked from the bravest of the troops which Ferdinand of Brunswick had led to victory at Creveld and Minden. But the excellent German commander, in this strange country, was no match for the astute Yankee on his own ground. Stealthily and leisurely, during the whole forenoon, the New England farmers marched around into Baum's rear. They did not march in military array, but in little squads, half a dozen at a time, dressed in their rustic blue frocks. There was nothing in their appearance which to a European veteran like Baum could seem at all soldier-like, and he thought that here at last were those blessed Tories, whom he had been taught to look out for, coming to place themselves

Battle of Bennington, Aug. 16, 1777.

behind him for protection. Early in the afternoon he was cruelly undeceived. For while 500 of these innocent creatures opened upon him a deadly fire in the rear and on both flanks, Stark, with 500 more, charged across the shallow stream and assailed him in front. The Indians instantly broke and fled screeching to the woods, while yet there was time for escape. The Germans stood their ground, and fought desperately; but thus attacked on all sides at once, they were soon thrown into disorder, and after a two hours' struggle, in which Baum was mortally wounded, they were all captured. At this moment, as the New England men began to scatter to the plunder of the German camp, the relieving force of Breymann came upon the scene; and the fortunes of the day might have been changed, had not Warner also arrived with his 500 fresh men in excellent order. A furious charge was made upon Breymann, who gave way, and retreated slowly from hill to hill, while parties of Americans kept pushing on to his rear to cut him off. By eight in the evening, when it had grown too dark to aim a gun, this second German force was entirely dispersed or captured. Breymann, with a mere corporal's guard of sixty or seventy men, escaped under cover of darkness, and reached the British camp in safety. Of the whole German force of 1,000 men, 207 had been killed and wounded, and more than 700 had been captured. Among the spoils of victory were 1,000 stand of arms, 1,000 dragoon swords, and the four field-pieces. Of the Americans 14 were killed and 42 wounded.

The invading force annihilated.

The news of this brilliant victory spread joy and hope throughout the land. Insubordination which had been crowned with such splendid success could not but be overlooked, and the gallant Stark was at once taken back into the army, and made a brigadier-general. Not least among the grounds of exultation was the fact that an army of yeomanry had not merely defeated, but annihilated, an army of the Brunswick regulars, with whose European reputation for bravery and discipline every man in the country was familiar. The bolder spirits began to ask the question why that which had been done to Baum and Breymann might not be done to Burgoyne's whole army; and in the excitement of this rising hope, reinforcements began to pour in faster and faster, both to Schuyler at Stillwater and to Lincoln at Manchester. On the other hand, Burgoyne at Fort Edward was fast losing heart, as dangers thickened around him. So far from securing his supplies of horses, wagons, and food by this stroke at Bennington, he had simply lost one seventh part of his available army, and he was now clearly in need of reinforcements as well as supplies. But no word had yet come from Sir William Howe, and the news from St. Leger was anything but encouraging. It is now time for us to turn westward and follow the wild fortunes of the second invading column.

Effect of the news; Burgoyne's enemies multiply.

About the middle of July, St. Leger had landed at Oswego, where he was joined by Sir John Johnson with his famous Tory regiment known as the

Royal Greens, and Colonel John Butler with his company of Tory rangers. Great efforts had been made by Johnson to secure the aid of the Iroquois tribes, but only with partial success. For once the Long House was fairly divided against itself, and the result of the present campaign did not redound to its future prosperity. The Mohawks, under their great chief Thayendanegea, better known as Joseph Brant, entered heartily into the British cause, and they were followed, though with less alacrity, by the Cayugas and Senecas; but the central tribe, the Onondagas, remained neutral. Under the influence of the missionary, Samuel Kirkland, the Oneidas and Tuscaroras actively aided the Americans, though they did not take the field. After duly arranging his motley force, which amounted to about 1,700 men, St. Leger advanced very cautiously through the woods, and sat down before Fort Stanwix on the 3d of August. This stronghold, which had been built in 1756, on the watershed between the Hudson and Lake Ontario, commanded the main line of traffic between New York and Upper Canada. The place was then on the very outskirts of civilization, and under the powerful influence of Johnson the Tory element was stronger here than in any other part of the state. Even here, however, the strength of the patriot party turned out to be much greater than had been supposed, and at the approach of the enemy the people began to rise in arms. In this part of New York there were many Germans, whose ancestors had come over to America during the horrors of

Advance of St. Leger upon Fort Stanwix.

the Thirty Years' War; and among these there was one stout patriot whose name shines conspicuously in the picturesque annals of the Revolution. General Nicholas Herkimer, commander of the militia of Tryon County, a veteran over sixty years of age, no sooner heard of St. Leger's approach than he started out to the rescue of Fort Stanwix; and by the 5th of August he had reached Oriskany, about eight miles distant, at the head of 800 men. The garrison of the fort, 600 in number, under Colonel Peter Gansevoort, had already laughed to scorn St. Leger's summons to surrender, when, on the morning of the 6th, they heard a distant firing to the eastward, which they could not account for. The mystery was explained when three friendly messengers floundered through a dangerous swamp into the fort, and told them of Herkimer's approach and of his purpose. The plan was to overwhelm St. Leger by a concerted attack in front and rear. The garrison was to make a furious sortie, while Herkimer, advancing through the forest, was to fall suddenly upon the enemy from behind; and thus it was hoped that his army might be crushed or captured at a single blow. To insure completeness of coöperation, Colonel Gansevoort was to fire three guns immediately upon receiving the message, and upon hearing this signal Herkimer would begin his march from Oriskany. Gansevoort would then make such demonstrations as to keep the whole attention of the enemy concentrated upon the fort, and thus guard Herkimer against a surprise by the way,

Herkimer marches against him.

Herkimer's plan.

until, after the proper interval of time, the garrison should sally forth in full force.

In this bold scheme everything depended upon absolute coördination in time. Herkimer had dispatched his messengers so early on the evening of the 5th that they ought to have reached the fort by three o'clock the next morning, and at about that time he began listening for the signal-guns. But through some unexplained delay it was nearly eleven in the forenoon when the messengers reached the fort, as just described. Meanwhile, as hour after hour passed by, and no signal-guns were heard by Herkimer's men, they grew impatient, and insisted upon going ahead, without regard to the preconcerted plan. Much unseemly wrangling ensued, in which Herkimer was called a coward and accused of being a Tory at heart, until, stung by these taunts, the brave old man at length gave way, and at about nine o'clock the forward march was resumed. At this time his tardy messengers still lacked two hours of reaching the fort, but St. Leger's Indian scouts had already discovered and reported the approach of the American force, and a strong detachment of Johnson's Greens under Major Watts, together with Brant and his Mohawks, had been sent out to intercept them.

Failure of the plan.

About two miles west of Oriskany the road was crossed by a deep semicircular ravine, concave toward the east. The bottom of this ravine was a swamp, across which the road was carried by a causeway of logs, and the steep banks on either side were thickly

Thayendanegea prepares an ambuscade.

covered with trees and underbrush. The practised eye of Thayendanegea at once perceived the rare advantage of such a position, and an ambuscade was soon prepared with a skill as deadly as that which once had wrecked the proud army of Braddock. But this time it was a meeting of Greek with Greek, and the wiles of the savage chief were foiled by a desperate valour which nothing could overcome. By ten o'clock the main body of Herkimer's army had descended into the ravine, followed by the wagons, while the rear-guard was still on the rising ground behind. At this moment they were greeted by a murderous volley from either side, while Johnson's Greens came charging down upon them in front, and the Indians, with frightful yells, swarmed in behind and cut off the rear-guard, which was thus obliged to retreat to save itself. For a moment the main body was thrown into confusion, but it soon rallied and formed itself in a circle, which neither bayonet charges nor musket fire could break or penetrate. The scene which ensued was one of the most infernal that the history of savage warfare has ever witnessed. The dark ravine was filled with a mass of fifteen hundred human beings, screaming and cursing, slipping in the mire, pushing and struggling, seizing each other's throats, stabbing, shooting, and dashing out brains. Bodies of neighbours were afterwards found lying in the bog, where they had gone down in a death-grapple, their cold hands still grasping the knives plunged in each other's hearts.

Battle of Oriskany, Aug. 6, 1777.

Early in the fight a musket-ball slew Herkimer's

horse, and shattered his own leg just below the knee; but the old hero, nothing daunted, and bating nothing of his coolness in the midst of the horrid struggle, had the saddle taken from his dead horse and placed at the foot of a great beech-tree, where, taking his seat and lighting his pipe, he continued shouting his orders in a stentorian voice and directing the progress of the battle. Nature presently enhanced the lurid horror of the scene. The heat of the August morning had been intolerable, and black thunder-clouds, overhanging the deep ravine at the beginning of the action, had enveloped it in a darkness like that of night. Now the rain came pouring in torrents, while gusts of wind howled through the tree-tops, and sheets of lightning flashed in quick succession, with a continuous roar of thunder that drowned the noise of the fray. The wet rifles could no longer be fired, but hatchet, knife, and bayonet carried on the work of butchery, until, after more than five hundred men had been killed or wounded, the Indians gave way and fled in all directions, and the Tory soldiers, disconcerted, began to retreat up the western road, while the patriot army, remaining in possession of the hard-won field, felt itself too weak to pursue them.

Retreat of the Tories.

At this moment, as the storm cleared away and long rays of sunshine began flickering through the wet leaves, the sound of the three signal-guns came booming through the air, and presently a sharp crackling of musketry was heard from the direction of Fort Stanwix. Startled by this ominous sound, the Tories made all possible haste to join their own army, while the patriots, bearing their wounded on litters of green boughs, returned in sad procession to Oriskany. With their commander helpless and more than one third of their number slain or disabled, they were in no condition to engage in a fresh conflict, and unwillingly confessed that the garrison of Fort Stanwix must be left to do its part of the work alone. Upon the arrival of the messengers, Colonel Gansevoort had at once taken in the whole situation. He understood the mysterious firing in the forest, saw that Herkimer must have been prematurely attacked, and ordered his sortie instantly, to serve as a diversion. The sortie was a brilliant success. Sir John Johnson, with his Tories and Indians, was completely routed and driven across the river. Colonel Marinus Willett took possession of his camp, and held it while seven wagons were three times loaded with spoil and sent to be unloaded in the fort. Among all this spoil, together with abundance of food and drink, blankets and clothes, tools and ammunition, the victors captured five British standards, and all Johnson's papers, maps, and memoranda, containing full instructions for the projected campaign. After this useful exploit, Colonel Willett returned to the fort and hoisted the captured British standards, while over them he raised an uncouth flag, intended to represent the American stars and stripes, which Congress had adopted in June as the national banner. This rude flag, hastily extemporized out of a white shirt, an old blue jacket, and some strips of red

Retreat of the patriot army.

Colonel Willett's sortie.

First hoisting of the stars and stripes.

cloth from the petticoat of a soldier's wife, was the first American flag with stars and stripes that was ever hoisted, and it was first flung to the breeze on the memorable day of Oriskany, August 6, 1777.

Of all the battles of the Revolution, this was perhaps the most obstinate and murderous. Each side seems to have lost not less than one third of its whole number; and of those lost, nearly all were killed, as it was largely a hand-to-hand struggle, like the battles of ancient times, and no quarter was given on either side. The number of surviving wounded, who were carried back to Oriskany, does not seem to have exceeded forty.

Death of Herkimer.

Among these was the indomitable Herkimer, whose shattered leg was so unskilfully treated that he died a few days later, sitting in bed propped by pillows, calmly smoking his Dutch pipe and reading his Bible at the thirty-eighth Psalm.

For some little time no one could tell exactly how the results of this fierce and disorderly day were to be regarded. Both sides claimed a victory, and St. Leger vainly tried to scare the garrison by the story that their comrades had been destroyed in the forest. But in its effects upon the campaign, Oriskany was for the Americans a success, though an incomplete one. St. Leger was not crushed, but he was badly crippled. The sacking of Johnson's camp injured his prestige in the neighbourhood, and the Indian allies, who had lost more than a hundred of their best warriors on that fatal morning, grew daily more sullen and refractory, until their strange behaviour came to be a fresh source of anxiety to the British commander. While he was pushing on the siege as well as he could, a force of 1,200 troops, under Arnold, was marching up the Mohawk valley to complete his discomfiture.

As soon as he had heard the news of the fall of Ticonderoga, Washington had dispatched Arnold to render such assistance as he could to the northern army, and Arnold had accordingly arrived at Schuyler's headquarters about three weeks ago. Before leaving Philadelphia, he had appealed to Congress to restore him to his former rank relatively to the five junior officers who had been promoted over him, and he had just learned that Congress had refused the request. At this moment, Colonel Willett and another officer, after a perilous journey through the wilderness, arrived at Schuyler's headquarters, and, bringing the news of Oriskany, begged that a force might be sent to raise the siege of Fort Stanwix. Schuyler understood the importance of rescuing the stronghold and its brave garrison, and called a council of war; but he was bitterly opposed by his officers, one of whom presently said to another, in an audible whisper, "He only wants to weaken the army!" At this vile insinuation, the indignant general set his teeth so hard as to bite through the stem of the pipe he was smoking, which fell on the floor and was smashed. "Enough!" he cried. "I assume the whole responsibility. Where is the brigadier who will go?" The brigadiers all sat in sullen silence; but Arnold, who had been brooding over his pri-

Arnold arrives at Schuyler's camp,

vate grievances, suddenly jumped up. "Here!" said he. "Washington sent me here to make myself useful: I will go." The commander gratefully seized him by the hand, and the drum beat for volunteers. Arnold's unpopularity in New England was mainly with the politicians. It did not extend to the common soldiers, who admired his impulsive bravery and had unbounded faith in his resources as a leader. Accordingly, 1,200 Massachusetts men were easily enlisted in the course of the next forenoon, and the expedition started up the Mohawk valley. Arnold pushed on with characteristic energy, but the natural difficulties of the road were such that after a week of hard work he had only reached the German Flats, where he was still more than twenty miles from Fort Stanwix. Believing that no time should be lost, and that everything should be done to encourage the garrison and dishearten the enemy, he had recourse to a stratagem, which succeeded beyond his utmost anticipation. A party of Tory spies had just been arrested in the neighbourhood, and among them was a certain Yan Yost Cuyler, a queer, half-witted fellow, not devoid of cunning, whom the Indians regarded with that mysterious awe with which fools and lunatics are wont to inspire them, as creatures possessed with a devil. Yan Yost was summarily condemned to death, and his brother and gypsy-like mother, in wild alarm, hastened to the camp, to plead for his life. Arnold for a while was inexorable, but presently offered to pardon the culprit on condition that he should go and spread a panic in the camp

and volunteers to relieve Fort Stanwix.

of St. Leger. Yan Yost joyfully consented, and started off forthwith, while his brother was detained as a hostage, to be hanged in case of his failure. To make the matter still surer, some friendly Oneidas were sent along to keep an eye upon him and act in concert with him. Next day, St. Leger's scouts, as they stole through the forest, began to hear rumours that Burgoyne had been totally defeated, and that a great American army was coming up the valley of the Mohawk. They carried back these rumours to the camp, and toward evening, while officers and soldiers were standing about in anxious consultation, Yan Yost came running in, with a dozen bullet-holes in his coat and terror in his face, and said that he had barely escaped with his life from the resistless American host which was close at hand. As many knew him for a Tory, his tale found ready belief, and when interrogated as to the numbers of the advancing host he gave a warning frown, and pointed significantly to the countless leaves that fluttered on the branches overhead. Nothing more was needed to complete the panic. It was in vain that Johnson and St. Leger exhorted and threatened the Indian allies. Already disaffected, they now began to desert by scores, while some, breaking open the camp chests, drank rum till they were drunk, and began to assault the soldiers. All night long the camp was a perfect Pandemonium. The riot extended to the Tories, and by noon of the next day St. Leger took to flight and his whole army was dispersed. All the tents, artillery, and stores fell into

Yan Yost Cuyler.

Flight of St. Leger, Aug. 22.

the hands of the Americans. The garrison, sallying forth, pursued St. Leger for a while, but the faithless Indians, enjoying his discomfiture, and willing to curry favour with the stronger party, kept up the chase nearly all the way to Oswego; laying ambushes every night, and diligently murdering the stragglers, until hardly a remnant of an army was left to embark with its crest-fallen leader for Montreal.

The news of this catastrophe reached Burgoyne before he had had time to recover from the news of the disaster at Bennington. Burgoyne's situation was now becoming critical. Lincoln, with a strong force of militia, was hovering in his rear, while the main army before him was gaining in numbers day by day. Putnam had just sent up reinforcements from the Highlands; Washington had sent Morgan with 500 sharpshooters; and Arnold was hurrying back from Fort Stanwix. Not a word had come from Sir William Howe, and it daily grew more difficult to get provisions.

Burgoyne's dangerous situation.

Just at this time, when everything was in readiness for the final catastrophe, General Gates arrived from Philadelphia, to take command of the northern army, and reap the glory earned by other men. On the first day of August, before the first alarm occasioned by Burgoyne's advance had subsided, Congress had yielded to the pressure of Schuyler's enemies, and removed him from his command; and on the following day Gates was appointed to take his place. Congress was led to take this step

Schuyler superseded by Gates, Aug. 2.

through the belief that the personal hatred felt toward Schuyler by many of the New England people would prevent the enlisting of militia to support him. The events of the next fortnight showed that in this fear Congress was quite mistaken. There can now be no doubt that the appointment of the incompetent Gates was a serious blunder, which might have ruined the campaign, and did in the end occasion much trouble, both for Congress and for Washington. Schuyler received the unwelcome news with the noble unselfishness which always characterized him. At no time did he show more zeal and diligence than during his last week of command; and on turning over the army to General Gates he cordially offered his aid, whether by counsel or action, in whatever capacity his successor might see fit to suggest. But so far from accepting this offer, Gates treated him with contumely, and would not even invite him to attend his first council of war. Such silly behaviour called forth sharp criticisms from discerning people. "The new commander-in-chief of the northern department," said Gouverneur Morris, "may, if he please, neglect to ask or disdain to receive advice; but those who know him will, I am sure, be convinced that he needs it."

Position of the two armies, Aug. 19–Sept. 12.

When Gates thus took command of the northern army, it was stationed along the western bank of the Hudson, from Stillwater down to Halfmoon, at the mouth of the Mohawk, while Burgoyne's troops were encamped along the eastern bank, some thirty miles higher up, from Fort Edward down to the

Battenkill. For the next three weeks no movements were made on either side; and we must now leave the two armies confronting each other in these two positions, while we turn our attention southward, and see what Sir William Howe was doing, and how it happened that Burgoyne had as yet heard nothing from him.

CHAPTER VII.

SARATOGA.

We have seen how, owing to the gross negligence of Lord George Germain, discretionary power had been left to Howe, while entirely taken away from Burgoyne. The latter had no choice but to move down the Hudson. The former was instructed to move up the Hudson, but at the same time was left free to depart from the strict letter of his instructions, should there be any manifest advantage in so doing. Nevertheless, the movement up the Hudson was so clearly prescribed by all sound military considerations that everybody wondered why Howe did not attempt it. Why he should have left his brother general in the lurch, and gone sailing off to Chesapeake Bay, was a mystery which no one was able to unravel, until some thirty years ago a document was discovered which has thrown much light upon the question. Here there steps again upon the scene that miserable intriguer, whose presence in the American army had so nearly wrecked the fortunes of the patriot cause, and who now, in captivity, proceeded to act the part of a doubly-dyed traitor. A marplot and mischief-maker from beginning to end, Charles Lee never failed to work injury to whichever party

Why Howe went to Chesapeake Bay.

Charles Lee in captivity.

his selfish vanity or craven fear inclined him for the moment to serve. We have seen how, on the day when he was captured and taken to the British camp, his first thought was for his personal safety, which he might well suppose to be in some jeopardy, since he had formerly held the rank of lieutenant-colonel in the British army. He was taken to New York and confined in the City Hall, where he was treated with ordinary courtesy; but there is no doubt that Sir William Howe looked upon him as a deserter, and was more than half inclined to hang him without ceremony. Fearing, however, as he said, that he might "fall into a law scrape," should he act too hastily, Sir William wrote home for instructions, and in reply was directed by Lord George Germain to send his prisoner to England for trial. In pursuance of this order, Lee had already been carried on board ship, when a letter from Washington put a stop to these proceedings. The letter informed General Howe that Washington held five Hessian field-officers as hostages for Lee's personal safety, and that all exchange of prisoners would be suspended until due assurance should be received that Lee was to be recognized as a prisoner of war. After reading this letter General Howe did not dare to send Lee to England for trial, for fear of possible evil consequences to the five Hessian officers, which might cause serious disaffection among the German troops. The king approved of this cautious behaviour, and so Lee was kept in New York, with his fate undecided, until it had become quite clear that neither arguments nor threats could avail one jot to shake

Washington's determination. When Lord George Germain had become convinced of this, he persuaded the reluctant king to yield the point; and Howe was accordingly instructed that Lee, although worthy of condign punishment, should be deemed a prisoner of war, and might be exchanged as such, whenever convenient.

All this discussion necessitated the exchange of several letters between London and New York, so that a whole year elapsed before the question was settled. It was not until December 12, 1777, that Howe received these final instructions. But Lee had not been idle all this time while his fate was in suspense. Hardly had the key been turned upon him in his rooms at the City Hall when he began his intrigues. First, he assured Lord Howe and his brother that he had always opposed the declaration of independence, and even now cherished hopes that, by a judiciously arranged interview with some of the delegates in Congress, he might persuade the misguided people of America to return to their old allegiance. Lord Howe, who always kept one hand on the olive-branch, eagerly caught at the suggestion, and permitted Lee to send a letter to Congress, urging that a committee be sent to confer with him, as he had "important communications to make." Could such a conference be brought about, he thought, his zeal for effecting a reconciliation would interest the Howes in his favour, and might save his precious neck. Congress, however, flatly refused to listen to the proposal, and then the wretch, without further ado, went over to the enemy, and

Treason of Charles Lee.

began to counsel with the British commanders how they might best subdue the Americans in the summer campaign. He went so far as to write out for the brothers Howe a plan of operations, giving them the advantage of what was supposed to be his intimate knowledge of the conditions of the case. This document the Howes did not care to show after the disastrous event of the campaign, and it remained hidden for eighty years, until it was found among the domestic archives of the Strachey family, at Sutton Court, in Somerset. The first Sir Henry Strachey was secretary to the Howes from 1775 to 1778. The document is in Lee's well-known handwriting, and is indorsed by Strachey as "Mr. Lee's plan, March 29, 1777." In this document Lee maintains that if the state of Maryland could be overawed, and the people of Virginia prevented from sending aid to Pennsylvania, then Philadelphia might be taken and held, and the operations of the "rebel government" paralyzed. The Tory party was known to be strong in Pennsylvania, and the circumstances under which Maryland had declared for independence, last of all the colonies save New York, were such as to make it seem probable that there also the loyalist feeling was very powerful. Lee did not hesitate to assert, as of his own personal knowledge, that the people of Maryland and Pennsylvania were nearly all loyalists, who only awaited the arrival of a British army in order to declare themselves. He therefore recommended that 14,000 men should drive Washington out of New Jersey and capture Philadelphia, while the remainder of

Howe's army, 4,000 in number, should go around by sea to Chesapeake Bay, and occupy Alexandria and Annapolis. From these points, if Lord Howe were to issue a proclamation of amnesty, the pacification of the "central colonies" might be effected in less than two months; and so confident of all this did the writer feel that he declared himself ready to "stake his life upon the issue," a remark which betrays, perhaps, what was uppermost in his mind throughout the whole proceeding. At the same time, he argued that offensive operations toward the north could not "answer any sort of purpose," since the northern provinces "are at present neither the seat of government, strength, nor politics; and the apprehensions from General Carleton's army will, I am confident, keep the New Englanders at home, or at least confine 'em to the east side the [Hudson] river."

It will be observed that this plan of Lee's was similar to that of Lord George Germain, in so far as it aimed at thrusting the British power like a wedge into the centre of the confederacy, and thus cutting asunder New England and Virginia, the two chief centres of the rebellion. But instead of aiming his blow at the Hudson river, Lee aims it at Philadelphia, as the "rebel capital;" and his reason for doing this shows how little he understood American affairs, and how strictly he viewed them in the light of his military experience in Europe. In European warfare it is customary to strike at the enemy's capital city, in order to get control of his whole system of administration; but that the pos-

Folly of moving upon Philadelphia, as the "rebel capital."

session of an enemy's capital is not always decisive the wars of Napoleon have most abundantly proved. The battles of Austerlitz in 1805 and Wagram in 1809 were fought by Napoleon after he had entered Vienna; it was not his acquisition of Berlin in 1806, but his victory at Friedland in the following summer, that completed the overthrow of Prussia; and where he had to contend against a strong and united national feeling, as in Spain and Russia, the possession of the capital did not help him in the least. Nevertheless, in European countries, where the systems of administration are highly centralized, it is usually advisable to move upon the enemy's capital. But to apply such a principle to Philadelphia in 1777 was the height of absurdity. Philadelphia had been selected for the meetings of the Continental Congress because of its geographical position. It was the most centrally situated of our large towns, but it was in no sense the centre of a vast administrative machinery. If taken by an enemy, it was only necessary for Congress to move to any other town, and everything would go on as before. As it was not an administrative, so neither was it a military centre. It commanded no great system of interior highways, and it was comparatively difficult to protect by the fleet. It might be argued, on the other hand, that because Philadelphia was the largest town in the United States, and possessed of a certain preëminence as the seat of Congress, the acquisition of it by the invaders would give them a certain moral advantage. It would help the Tory party, and discourage the patriots. Such a gain, however, would be trifling compared with the loss which might come from Howe's failure to coöperate with Burgoyne; and so the event most signally proved.

Just how far the Howes were persuaded by Lee's arguments must be a matter of inference. The course which they ultimately pursued, in close conformity with the suggestions of this remarkable document, was so disastrous to the British cause that the author might almost seem to have been intentionally luring them off on a false scent. One would gladly take so charitable a view of the matter, were it not both inconsistent with what we have already seen of Lee, and utterly negatived by his scandalous behaviour the following year, after his restoration to his command in the American army. We cannot doubt that Lee gave his advice in sober earnest. That considerable weight was attached to it is shown by a secret letter from Sir William Howe to Lord George Germain, dated the 2d of April, or four days after the date of Lee's extraordinary document. In this letter, Howe intimates for the first time that he has an expedition in mind which may modify the scheme for a joint campaign with the northern army along the line of the Hudson. To this suggestion Lord George replied on the 18th of May: "I trust that whatever you may meditate will be executed in time for you to coöperate with the army to proceed from Canada." It was a few days after this that Lord George, perhaps feeling a little uneasy about the matter, wrote that imperative order which lay in its pigeon-hole in London until all the damage was done.

Effect of Lee's advice.

With these data at our command, it becomes easy to comprehend General Howe's movements during the spring and summer. His first intention was to push across New Jersey with the great body of his army, and occupy Philadelphia; and since he had twice as many men as Washington, he might hope to do this in time to get back to the Hudson as soon as he was likely to be needed there. He began his march on the 12th of June, five days before Burgoyne's flotilla started southward on Lake Champlain. The enterprise did not seem hazardous, but Howe was completely foiled by Washington's superior strategy. Before the British commander had fairly begun to move, Washington, from various symptoms, divined his purpose, and coming down from his lair at Morristown, planted himself on the heights of Middlebrook, within ten miles of New Brunswick, close upon the flank of Howe's line of march. Such a position, occupied by 8,000 men under such a general, was equivalent to a fortress which it would not do for Howe to pass by and leave in his rear. But the position was so strong that to try to storm it would be to invite defeat. It remained to be seen what could be done by manœuvring. The British army of 18,000 men was concentrated at New Brunswick, with plenty of boats for crossing the Delaware river, when that obstacle should be reached. But the really insuperable obstacle was close at hand. A campaign of eighteen days ensued, consisting of wily marches and counter-marches, the result of which showed that Washington's advantage of position

Washington's masterly campaign in New Jersey, June, 1777.

could not be wrested from him. Howe could neither get by him nor outwit him, and was too prudent to attack him; and accordingly, on the last day of June, he abandoned his first plan, and evacuated New Jersey, taking his whole army over to Staten Island.

This campaign has attracted far less attention than it deserves, mainly, no doubt, because it contained no battles or other striking incidents. It was purely a series of strategic devices. But in point of military skill it was, perhaps, as remarkable as anything that Washington ever did, and it certainly occupies a cardinal position in the history of the overthrow of Burgoyne. For if Howe had been able to take Philadelphia early in the summer, it is difficult to see what could have prevented him from returning and ascending the Hudson, in accordance with the plan of the ministry. Now the month of June was gone, and Burgoyne was approaching Ticonderoga. Howe ought to have held himself in readiness to aid him, but he could not seem to get Philadelphia, the "rebel capital," out of his mind. His next plan coincided remarkably with the other half of Lee's scheme. He decided to go around to Philadelphia by sea, but he was slow in starting, and seems to have paused for a moment to watch the course of events at the north. He began early in July to put his men on board ship, but confided his plans to no one but Cornwallis and Grant; and his own army, as well as the Americans, believed that this show of going to sea was only a feint to disguise his real intention.

Uncertainty as to Howe's next movements.

Every one supposed that he would go up the Hudson. As soon as New Jersey was evacuated Washington moved back to Morristown, and threw his advance, under Sullivan, as far north as Pompton, so as to be ready to coöperate with Putnam in the Highlands, at a moment's notice. As soon as it became known that Ticonderoga had fallen, Washington, supposing that his adversary would do what a good general ought to do, advanced into the Ramapo Clove, a rugged defile in the Highlands, near Haverstraw, and actually sent the divisions of Sullivan and Stirling across the river to Peekskill.

All this while Howe kept moving some of his ships, now up the Hudson, now into the Sound, now off from Sandy Hook, so that people might doubt whether his destination were the Highlands, or Boston, or Philadelphia. Probably his own mind was not fully made up until after the news from Ticonderoga. Then, amid the general exultation, he seems to have concluded that Burgoyne would be able to take care of himself, at least with such coöperation as he might get from Sir Henry Clinton. In this mood he wrote to Burgoyne as follows: "I have . . . heard from the rebel army of your being in possession of Ticonderoga, which is a great event, carried without loss. . . . Washington is waiting our motions here, and has detached Sullivan with about 2,500 men, as I learn, to Albany. My intention is for Pennsylvania, where I expect to meet Washington; but if he goes to the northward, contrary to my expectations, and you can keep him at bay, be assured I

Howe's letter to Burgoyne.

shall soon be after him to relieve you. After your arrival at Albany, the movements of the enemy will guide yours; but my wishes are that the enemy be drove [*sic*] out of this province before any operation takes place in Connecticut. Sir Henry Clinton remains in the command here, and will act as occurrences may direct. Putnam is in the Highlands with about 4,000 men. Success be ever with you." This letter, which was written on very narrow strips of thin paper, and conveyed in a quill, did not reach Burgoyne till the middle of September, when things wore a very different aspect from that which they wore in the middle of July. Nothing could better illustrate the rash, overconfident spirit in which Howe proceeded to carry out his southern scheme. A few days afterward he put to sea with the fleet of 228 sail, carrying an army of 18,000 men, while 7,000 were left in New York, under Sir Henry Clinton, to garrison the city and act according to circumstances. Just before sailing Howe wrote a letter to Burgoyne, stating that the destination of his fleet was Boston, and he artfully contrived that this letter should fall into Washington's hands. But Washington was a difficult person to hoodwink. On reading the letter he rightly inferred that Howe had gone southward. Accordingly, recalling Sullivan and Stirling to the west side of the Hudson, he set out for the Delaware, but proceeded very cautiously, lest Howe should suddenly retrace his course, and dart up the Hudson. To guard against such an emergency, he let Sullivan advance no farther than Morristown, and kept everything in readiness for an instant counter-march. In a letter of July 30th he writes, "Howe's in a manner abandoning Burgoyne is so unaccountable a matter that, till I am fully assured of it, *I cannot help casting my eyes continually behind me.*" Next day, learning that the fleet had arrived at the Capes of Delaware, he advanced to Germantown; but on the day after, when he heard that the fleet had put out to sea again, he suspected that the whole movement had been a feint. He believed that Howe would at once return to the Hudson, and immediately ordered Sullivan to counter-march, while he held himself ready to follow at a moment's notice. His best generals entertained the same opinion. "I cannot persuade myself," said Greene, "that General Burgoyne would dare to push with such rapidity towards Albany if he did not expect support from General Howe." A similar view of the military exigencies of the case was taken by the British officers, who, almost to a man, disapproved of the southward movement. They knew as well as Greene that, however fine a city Philadelphia might be, it was "an object of far less military importance than the Hudson river."

Comments of Washington and Greene.

No wonder that the American generals were wide of the mark in their conjectures, for the folly of Howe's movements after reaching the mouth of the Delaware was quite beyond credence, and would be inexplicable to-day except as the result of the wild advice of the marplot Lee. Howe alleged as his reason for turning away from the Delaware that there were obstructions in the river and forts to pass, and

Howe's alleged reason trumped up and worthless.

accordingly he thought it best to go around by way of Chesapeake Bay, and land his army at Elkton. Now he might easily have gone a little way up the Delaware river without encountering any obstructions whatever, and landed his troops at a point only thirteen miles east of Elkton. Instead of attempting this, he wasted twenty-four days in a voyage of four hundred miles, mostly against head-winds, in order to reach the same point! No sensible antagonist could be expected to understand such eccentric behaviour. No wonder that, after it had become clear that the fleet had gone southward, Washington should have supposed an attack on Charleston to be intended. A council of war on the 21st decided that this must be the case, and since an overland march of seven hundred miles could not be accomplished in time to prevent such an attack, it was decided to go back to New York, and operate against Sir Henry Clinton. But before this decision was acted on Howe appeared at the head of Chesapeake Bay, where he landed his forces at Elkton. It was now the 25th of August, — nine days after the battle of Bennington and three days after the flight of St. Leger. Since entering Chesapeake Bay, Howe had received Lord George Germain's letter of May 18th, telling him that whatever he had to do ought to be done in time for him to coöperate with Burgoyne. Now Burgoyne's situation had become dangerous, and here was Howe at Elkton, fifty miles southwest of Philadelphia, with Washington's army in front of him, and more than three hundred miles away from Burgoyne!

Burgoyne's fate practically decided.

On hearing of Howe's arrival at the head of Chesapeake Bay, Washington had advanced as far as Wilmington to meet him. The first proceeding of the British general, on landing at Elkton, was to issue his proclamation of amnesty; but it did not bring him many recruits. A counter-proclamation, drawn up by Luther Martin, sufficed to neutralize it. Though there were many people in the neighbourhood who cared little for the cause of independence, there were but few who sympathized with the invaders enough to render them any valuable assistance. It was through a country indifferent, perhaps, but not friendly in feeling, that the British army cautiously pushed its way northward for a fortnight, until it reached the village of Kennett Square, six miles west of the Brandywine Creek, behind which Washington had planted himself to oppose its progress.

Washington's reasons for offering battle.

The time had arrived when Washington felt it necessary to offer battle, even though such a step might not be justified from purely military reasons. The people were weary of a Fabian policy which they did not comprehend, and Washington saw that even if he were defeated, the moral effect upon the country would not be so bad as if he were to abandon Philadelphia without a blow. A victory he was hardly entitled to expect, since he had but 11,000 men against Howe's 18,000, and since the British were still greatly superior in equipment and discipline. Under these circumstances Washington chose his ground with his usual sagacity, and took possession of it by a swift and masterly movement. The

Brandywine Creek ran directly athwart Howe's line of march to Philadelphia. Though large enough to serve as a military obstacle, — in England it would be called a river, — it was crossed by numerous fords, of which the principal one, Chadd's Ford, lay in Howe's way. Washington placed the centre of his army just behind Chadd's Ford and across the road. His centre was defended in front by a corps of artillery under Wayne, while Greene, on some high ground in the rear, was stationed as a reserve. Below Chadd's Ford, the Brandywine becomes a roaring torrent, shut in between steep, high cliffs, so that the American left, resting upon these natural defences, was sufficiently guarded by the Pennsylvania militia under Armstrong. The right wing, stretching two miles up the stream, into an uneven and thickly wooded country, was commanded by Sullivan.

He chooses a very strong position.

This was a very strong position. On the left it was practically inaccessible. To try storming it in front would be a doubtful experiment, sure to result in terrible loss of life. The only weak point was the right, which could be taken in flank by a long circuitous march through the woods. Accordingly, on the morning of the 11th of September, the British right wing, under Knyphausen, began skirmishing and occupying Washington's attention at Chadd's Ford; while the left column, under the energetic Cornwallis, marched up the Lancaster road, crossed the forks of the Brandywine, and turned southward toward Birmingham church, with the

Battle of the Brandywine, Sept. 11, 1777.

intention of striking the rear of the American right wing. It was similar to the flanking movement which had been tried so successfully at the battle of Long Island, a year before. It was quite like the splendid movement of Stonewall Jackson at Chancellorsville, eighty-five years afterward. In Howe's time such flanking marches were eminently fashionable. It was in this way that the great Frederick had won some of his most astonishing victories. They were, nevertheless, then as always, dangerous expedients, as the stupendous overthrow of the Austro-Russian army at Austerlitz was by and by to show. There is always a serious chance that the tables may be turned. Such flanking movements are comparatively safe, however, when the attacking army greatly outnumbers the army attacked, as at the Brandywine. But in all cases the chief element in their success is secrecy; above all things, the party attacked must be kept in the dark.

These points are admirably illustrated in the battle of the Brandywine. The danger of a flank attack upon his right wing was well understood by Washington; and as soon as he heard that Cornwallis was marching up the Lancaster road, he considered the feasibleness of doing what Frederick would probably have done, — of crossing quickly at Chadd's and Brinton's fords, in full force, and crushing Knyphausen's division. This he could doubtless have accomplished, had he been so fortunate as to have inherited an army trained by the father of Frederick the Great. But Washington's army was not yet well trained, and its

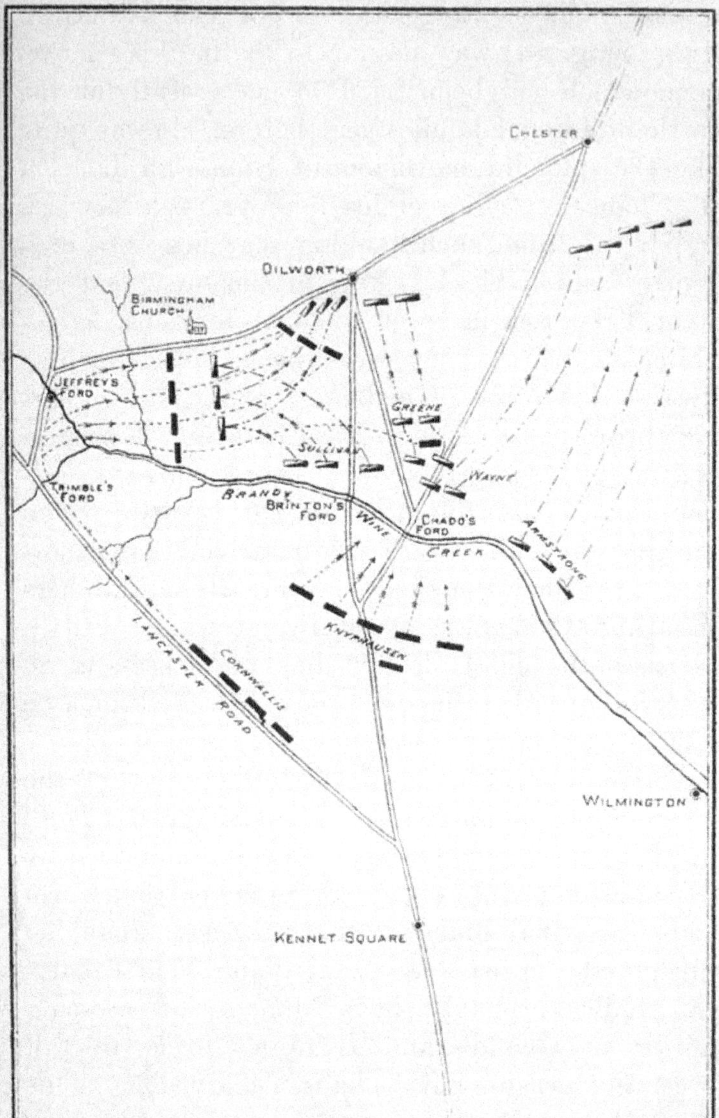

BATTLE OF THE BRANDYWINE

SEPTEMBER 11, 1777

numerical inferiority was such that Knyphausen's division might of itself be regarded as a fair match for it. The British movement was, therefore, well considered, and it was doubtless right that Washington did not return the offensive by crossing the creek. Moreover, the organization of his staff was far from complete. He was puzzled by conflicting reports as to the enemy's movements. While considering the question of throwing his whole force against Knyphausen, he was stopped by a false report that Cornwallis was *not* moving upon his flank. So great was the delay in getting intelligence that Cornwallis had accomplished his long march of eighteen miles, and was approaching Birmingham church, before it was well known where he was. Nevertheless, his intention of dealing a death-blow to the American army was forestalled and partially checked. Before he had reached our right wing, Washington had ordered Sullivan to form a new front and advance toward Birmingham church. Owing to the imperfect discipline of the troops, Sullivan executed the movement rather clumsily, but enough was accomplished to save the army from rout. In the obstinate and murderous fight which ensued near Birmingham church between Cornwallis and Sullivan, the latter was at length slowly pushed back in the direction of Dilworth. To save the army from being broken in two, it was now necessary for the centre to retreat upon Chester by way of Dilworth, and this movement was accomplished by Greene with consummate skill. It was now possible for Knyphausen to advance across Chadd's Ford against Wayne's position; and he did so, aided by the right wing of Cornwallis's division, which, instead of joining in the oblique pursuit toward Dilworth, kept straight onward, and came down upon Wayne's rear. Nothing was left for Wayne and Armstrong but to retreat and join the rest of the army at Chester, and so the battle of the Brandywine came to an end.

This famous battle was admirably conducted on both sides. The risk assumed in the long flanking march of Cornwallis was fully justified. The poor organization of the American army was of course well known to the British commanders, and they took advantage of the fact. Had they been dealing with an organization as efficient as their own, their course would have been foolhardy. On the other hand, when we consider the relative strength of the two armies, it is clear that the bold move of Cornwallis ought not simply to have won the field of battle. It ought to have annihilated the American army, had not its worst consequences been averted by Washington's promptness, aided by Sullivan's obstinate bravery and Greene's masterly conduct of the retreat upon Dilworth. As it was, the American soldiers came out of the fight in good order. Nothing could be more absurd than the careless statement, so often made, that the Americans were "routed" at the Brandywine. Their organization was preserved, and at Chester, next day, they were as ready for fight as ever. They had exacted from the enemy a round price for the victory. The American loss was a little more than 1,000, incurred chiefly in Sullivan's gallant struggle; rolls afterward captured at Germantown showed that the British loss considerably exceeded that figure.

So far as the possession of Philadelphia was concerned, the British victory was decisive. When the news came, next morning, that the army had retreated upon Chester, there was great consternation in the "rebel capital." Some timid people left their homes, and sought refuge in the mountains. Congress fled to Lancaster, first clothing Washington for sixty days with the same extraordinary powers which had been granted him the year before. Yet there was no need of such unseemly haste, for Washington detained the victorious enemy a fortnight on the march of only twenty-six miles; a feat which not even Napoleon could have performed with an army that had just been "routed." He had now heard of Stark's victory and St. Leger's flight, and his letters show how clearly he foresaw Burgoyne's inevitable fate, provided Howe could be kept away from him. To keep Howe's whole force employed near Philadelphia as long as possible was of the utmost importance. Accordingly, during the fortnight following the battle of the Brandywine, every day saw manœuvres or skirmishes, in one of which General Wayne was defeated by Sir Charles Grey, with a loss of three hundred men. On the 26th, while Howe established his headquarters at Germantown, Cornwallis entered Philadelphia in triumph, marching with bands of music and flying colours, and all the troops decked out in their finest scarlet array.

Washington's skill in detaining the enemy.

The British enter Philadelphia, Sept. 26.

Having got possession of the "rebel capital," the question now arose whether it would be possible to hold it through the winter. The Delaware river, below the city, had been carefully obstructed by *chevaux-de-frise*, which were guarded by two strong fortresses, — Fort Mifflin on an island in mid-stream, and Fort Mercer on the Jersey shore. The river was here about two miles in width, but it was impossible for ships to pass until the forts should have been reduced. About the first of October, after a rough return voyage of four hundred miles, Lord Howe's fleet appeared at the mouth of the Delaware. It was absolutely necessary to gain control of the river, in order that the city might get supplies by sea; for so long as Washington's army remained unbroken, the Americans were quite able to cut off all supplies by land. Sir William Howe, therefore, threw a portion of his forces across the river, to aid his brother in reducing the forts. The quick eye of Washington now saw an opportunity for attacking the main British army, while thus temporarily weakened; and he forthwith planned a brilliant battle, which was fated to be lost, at the very moment of victory, by an extraordinary accident.

Significance of Forts Mercer and Mifflin.

The village of Germantown, by the bank of the Schuylkill river, was then separated from Philadelphia by about six miles of open country. The village consisted chiefly of a single street, about two miles in length, with stone houses on either side, standing about a hundred yards apart from each other, and sur-

The situation at Germantown.

rounded by gardens and orchards. Near the upper end of the street, in the midst of ornamental shrubbery, vases, and statues, arranged in a French style of landscape gardening, stood the massively built house of Benjamin Chew, formerly Chief Justice of Pennsylvania. About a mile below, at the Market House, the main street was crossed at right angles by the Old School Lane. Beside the main street, running over Chestnut Hill, the village was approached from the northward by three roads. The Monatawny road ran down by the bank of the Schuylkill, and, crossing the Old School Lane, bore on toward Philadelphia. The Limekiln road, coming from the northeast, became continuous with the Old School Lane. The Old York road, still further eastward, joined the main street at the Rising Sun tavern, about two miles below the Market House.

The British army lay encamped just behind the Old School Lane, in the lower part of the village: the left wing, under Knyphausen, to the west of the main street; the right, under Grant, to the east. A strong detachment of *chasseurs*, under Sir Charles Grey, covered the left wing. About a mile in advance of the army, Colonel Musgrave's regiment lay in a field opposite Judge Chew's house; and yet a mile farther forward a battalion of light infantry was stationed on the slight eminence known as Mount Airy, where a small battery commanded the road to the north.

Washington's plan of attack seems to have contemplated nothing less than the destruction or capture of the British army. His forces were to advance from the north by all four roads at once, and converge upon the British at the Market House. The American right wing, under Sullivan, and consisting of Sullivan's own brigade, with those of Conway, Wayne, Maxwell, and Nash, was to march down the main street, overwhelm the advanced parties of the British, and engage their left wing in front; while Armstrong, with the Pennsylvania militia, was to move down the Monatawny road, and take the same wing in flank. The American left wing, commanded by Greene, was also to proceed in two columns. Greene, with his own brigade, supported by Stephen and McDougal, was to march down the Limekiln road, and assail the British right wing in front and in flank; while Smallwood and Forman, coming down the Old York road, were to strike the same wing in the rear. The flank attack upon the British left, entrusted as it was to militia, was intended merely as a demonstration. The attack upon their right, conducted by more than half of the American army, including its best troops, was intended to crush that wing, and folding back the whole British army upon the Schuylkill river, compel it to surrender.

Washington's audacious plan.

Considering that the Americans had not even yet a superiority in numbers, this was a most audacious plan. No better instance could be given of the spirit of wild and venturous daring which was as conspicuous in Washington as his cautious vigilance, whenever any fit occasion arose for displaying it. The scheme came surprisingly near to success; so near as to redeem it from the imputa-

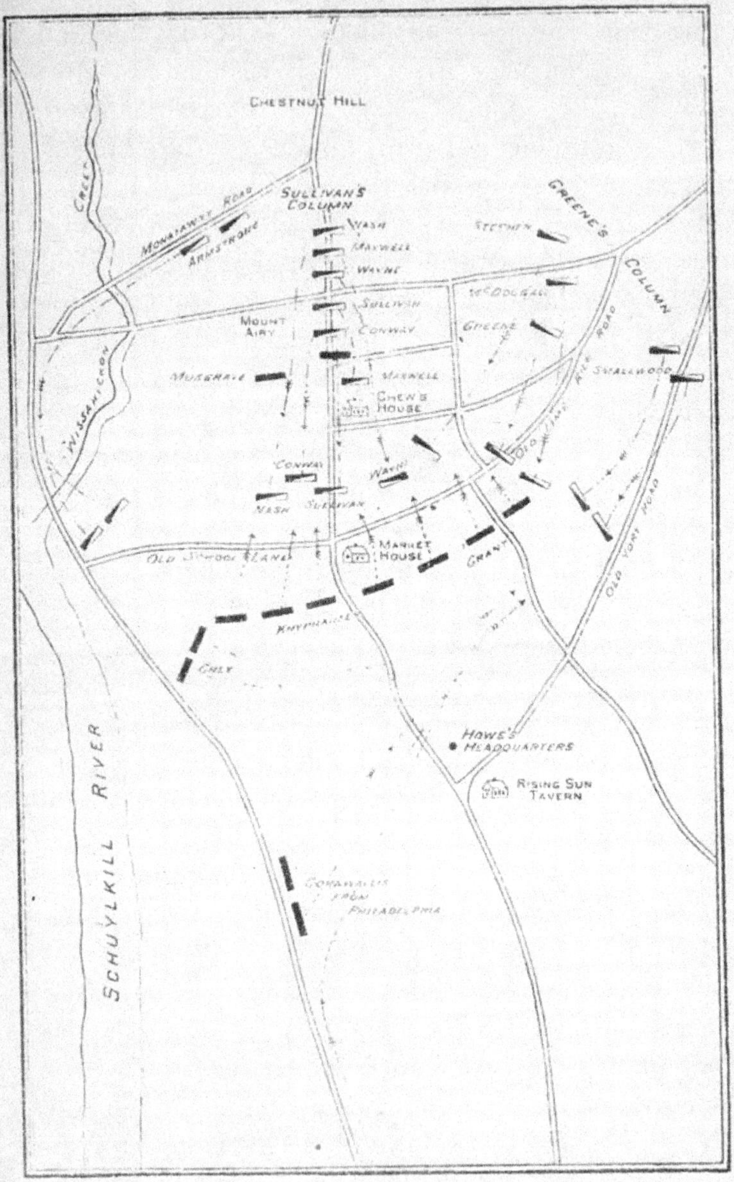

BATTLE OF GERMANTOWN

OCTOBER 4, 1777

tion of foolhardiness, and to show that here, as in all Washington's military movements, cool judgment went along with fiery dash. At seven in the evening of the 3d of October, the night march upon Germantown began, Washington accompanying Sullivan's column. At sunrise a heavy fog came up, and the darkness went on increasing. Soon after the hour of daybreak the light infantry upon Mount Airy were surprised and routed, and the battery was captured. Musgrave was next overwhelmed by the heavy American column; but he, with a small force, took refuge in Judge Chew's house, and set up a brisk fire from the windows. The Americans opened an artillery-fire upon the house, but its stone walls were too solid to be beaten down by the three-pound and six-pound field-pieces of that day; and so Maxwell's brigade was left behind to besiege the house, while the rest of the column rushed on down the street. The chief effect of this incident was to warn the enemy, while retarding and somewhat weakening the American charge. Nevertheless, the fury of the attack was such as to disconcert Knyphausen's veterans, and the British left wing slowly gave way before Sullivan. At this moment, Greene, who had also been delayed, attacked the right wing with such vigour as presently to force it back toward the Market House. The British ranks were falling into confusion, and Smallwood's column had already arrived upon their right flank, when the accident occurred which changed the fortunes of the day. From the beginning the dense fog had been a source of confu-

Battle of Germantown, Oct. 4.

sion to both armies, and had seriously interfered with the solidity of the American advance. Now, as Stephen's brigade, on the right of Greene's column, came into the village, the heavy firing at Judge Chew's seems to have caused him to diverge more and more to the west, in the belief that there was the thick of the battle. At the same time, Wayne, in driving the enemy before him, had swayed somewhat to the east, so that his brigade stood almost directly in the line of Stephen's progress. In this position he was attacked by Stephen, who mistook him for the enemy. This lamentable blunder instantly ruined the battle. Wayne's men, thus fiercely attacked in the rear, and struggling to extricate themselves, were thrown upon the left flank of Sullivan's brigade, and a panic suddenly ran through the army. The confusion grew worse and worse, till a general retreat began, and Grey, who had come up to support the crumbling right wing of the British, was now able to lead in the pursuit of the Americans. He was joined by Cornwallis, who had sprung from his bed in Philadelphia at the first sound of the cannon, and had brought up two battalions with him at double-quick. But the panic had subsided almost as soon as the golden moment of victory was lost, and the retreat was conducted in excellent order. One regiment in Greene's column was surrounded and captured, but the army brought away all its cannon and wounded, with several cannon taken from the enemy. The loss of the Americans in killed and wounded was 673, and the loss of the British was 535.

The fog which enshrouded the village of Germantown on that eventful morning has been hardly less confusing to historians than it was to the armies engaged. The reports of different observers conflicted in many details, and particularly as to the immediate occasion of the fatal panic. The best accounts agree, however, that the entanglement of Stephen with Wayne was chiefly responsible for the disaster. It was charged against Stephen that he had taken too many pulls at his canteen on the long, damp night march, and he was tried by court-martial, and dismissed from the service. The chagrin of the Americans at losing the prize so nearly grasped was profound. The total rout of Howe, coming at the same time with the surrender of Burgoyne, would probably have been too much for Lord North's ministry to bear, and might have brought the war to a sudden close. As it was, the British took an undue amount of comfort in the acquisition of Philadelphia, though so long as Washington's army remained defiant it was of small military value to them. On the other hand, the genius and audacity shown by Washington, in thus planning and so nearly accomplishing the ruin of the British army only three weeks after the defeat at the Brandywine, produced a profound impression upon military critics in Europe. Frederick of Prussia saw that presently, when American soldiers should come to be disciplined veterans, they would become a very formidable instrument in the hands of their great commander; and the French court, in making up its mind that the Americans would prove efficient allies, is said to have been influenced almost as much by the battle of Germantown as by the surrender of Burgoyne.

Howe captures Forts Mercer and Mifflin.

Having thus escaped the catastrophe which Washington had designed for him, the British commander was now able to put forth his utmost efforts for the capture of the forts on the Delaware. His utmost efforts were needed, for in the first attack on Fort Mercer, October 22, the Hessians were totally defeated, with the loss of Count Donop and 400 men, while the Americans lost but 37. But after a month of hard work, with the aid of 6,000 more men sent from New York by Clinton, both forts were reduced, and the command of the Delaware was wrested from the Americans. Another month of manœuvring and skirmishing followed, and then Washington took his army into winter-quarters at Valley Forge. The events which attended his sojourn in that natural stronghold belong to a later period of the war. We must now return to the upper waters of the Hudson, and show how the whole period, which may be most fitly described as a struggle for the control of the great central state of New York, was brought to an end by the complete and overwhelming victory of the Americans.

We have seen how it became impossible for Howe to act upon Lord George Germain's order, received in August, in Chesapeake Bay, and get back to the Hudson in time to be of any use to Burgoyne. We have also seen how critical was the situation in which the northern general was left, after the destruction of Baum and St. Leger, and the accumulation of New England yeomanry in his rear. Burgoyne now fully acknowledged the terrible mistake of the ministry in assuming that the resistance of the Americans was due to the machinations of a few wily demagogues, and that the people would hail the approach of the king's troops as deliverers. "The great bulk of the country," said he, "is undoubtedly with the Congress in principle and zeal, and their measures are executed with a secrecy and dispatch that are not to be equalled. . . . The Hampshire Grants, in particular, a country unpeopled and almost unknown last war, now abounds in the most active and most rebellious race on the continent, and hangs like a gathering storm upon my left." The situation had, indeed, become so alarming that it is hard to say what Burgoyne ought to have done. A retreat upon Ticonderoga would have been fraught with peril, while to cross the Hudson and advance upon Albany would be doing like Cortes, when he scuttled his ships. But Burgoyne was a man of chivalrous nature. He did not think it right or prudent to abandon Sir William Howe, whom he still supposed to be coming up the river to meet him. In a letter to Lord George Germain, written three days after the surrender, he says, "The difficulty of a retreat upon Canada was clearly foreseen, as was the dilemma, should the retreat be effected, of leaving at liberty such an army as General Gates's to operate against Sir William Howe. This consideration operated forcibly to determine me to

Burgoyne recognizes the fatal error of Germaine.

abide events as long as possible, and I reasoned thus: the expedition which I commanded was at first evidently intended to be *hazarded*; circumstances might require it should be *devoted*."

Influenced by these views, which were supported by all his generals except Riedesel, Burgoyne threw a bridge of boats across the Hudson, and passed over with his whole army on the 13th of September. The Americans had taken a strong position on Bemis Heights, where Kosciusko had skilfully fortified their camp with batteries and redoubts. Burgoyne felt that the time for desperate fighting had now come, and it seemed to him that the American position might be turned and carried by an attack upon its left flank. On the morning of the 19th, he advanced through the woods, with the centre of his army, toward the point where the Quaker road passes Bemis Heights. The right wing, under Fraser, proceeded somewhat more circuitously toward the same point, the plan being that they should join forces and strike the rear of the American camp, while Riedesel and Phillips, with the left wing and the artillery, marching down the river road, should assail it in front. Three heavy guns, announcing to the left wing the junction of Burgoyne and Fraser, were to give the signal for a general assault. American scouts, lurking among the upper branches of tall trees that grew on steep hillsides, presently caught glimpses of bright scarlet flitting through the green depths of the forest, while the long sunbeams that found their way through the foliage sent back

Nevertheless, he crosses the Hudson.

First battle at Freeman's Farm, Sept. 19; indecisive.

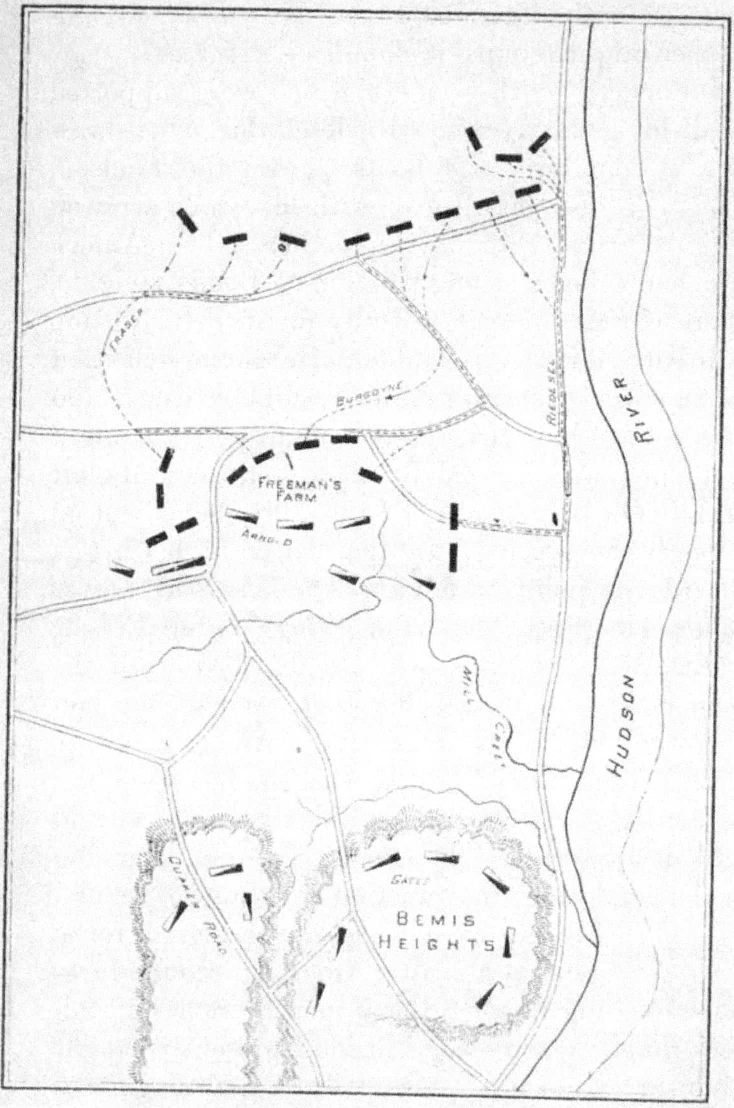

FIRST BATTLE AT FREEMAN'S FARM

SEPTEMBER 19, 1777

quick burning flashes from a thousand bayonets. By noon the course of the British march and their plan of attack had been fully deciphered, and the intelligence was carried to Arnold, who commanded the left wing of the American army. Gates appears to have been unwilling to let any of the forces descend from their strong position; but the fiery Arnold urged and implored, until he got permission to take Morgan's riflemen and Dearborn's infantry, and go forth to attack the enemy. Arnold's advance, under Morgan, first fell upon Burgoyne's advance, at Freeman's Farm, and checked its progress. Fraser then, hearing the musketry, turned eastward to the rescue, while Arnold, moving upon Fraser's left, sought to cut him asunder from Burgoyne. He seemed to be winning the day, when he was attacked in flank by Riedesel, who had hurried up from the river road. Arnold had already sent to Gates for reinforcements, which were refused him. Arnold maintained that this was a gross blunder on the part of the commanding general, and that with 2,000 more men he could now easily have crushed the British centre and defeated their army. In this opinion he was probably right, since even as it was he held his own, in a desperate fight, for two hours, until darkness put an end to the struggle. The losses on each side are variously estimated at from 600 to 1,000, or from one fifth to one fourth of the forces engaged, which indicates severe fighting. Arnold's command had numbered about 3,000, and he had been engaged, in the course of the afternoon, with at least 4,000 of Burgoyne's army; yet all this while some 11,000 Americans — most of the army in short — had been kept idle on Bemis Heights by the incompetent Gates. Burgoyne tried to console himself with the idea that he had won a victory, because his army slept that night at Freeman's Farm; but in his testimony given afterward before the House of Commons, he rightly maintained that his plan of attack had been utterly defeated by the bold and skilful tactics of "Mr." Arnold.

Quarrel between Gates and Arnold.

In the dispatches which he now sent to Congress, Gates took to himself all the credit of this affair, and did not even mention Arnold's name. The army, however, rang with praise of the fighting general, until Gates, who never could bear to hear any one but himself well spoken of, waxed wroth and revengeful. Arnold, moreover, freely blamed Gates for not supporting him, and for refusing to renew the battle on the next morning, while the enemy were still disconcerted. Arnold's warm friendship with Schuyler gave further offence to the commander; and three days after the battle he sought to wreak his spite by withdrawing Morgan's riflemen and Dearborn's light infantry from Arnold's division. A fierce quarrel ensued; and Gates told Arnold that as soon as Lincoln should arrive he would have no further use for him, and he might go back to Washington's camp as soon as he liked. Arnold, in a white rage, said he would go, and asked for a pass, which his enemy promptly gave him; but after receiving it, second thoughts prevented him from going. All the general officers except Lincoln — who seems to have refrained from unwillingness to give umbrage to a commander so high in the good graces of Massachusetts as Gates — united in signing a letter entreating Arnold to remain. He had been sent here by Washington to aid the northern army with all his might, and clearly it would be wrong to leave it now, on the eve of a decisive battle. So the proud, fiery soldier, smarting under an accumulation of injuries, made up his mind once more to swallow the affront, and wait for a chance to make himself useful. He stayed in his quarters, awaiting the day of battle, though it was not clear how far he was entitled, under the circumstances, to exercise command, and Gates took no more notice of him than if he had been a dog.

Nothing more was done for eighteen days. Just before the crossing of the Hudson by the northern army, Sir Henry Clinton, acting "as circumstances may direct," had planned an expedition up the river in aid of it; and Burgoyne, hearing of this the day after the battle at Freeman's Farm, thought it best to wait a while before undertaking another assault upon the American lines. But things were swiftly coming to such a pass that it would not do to wait. On the 21st, news came to the British camp that a detachment of Lincoln's troops had laid siege to Ticonderoga, and, while holding the garrison in check, had captured several ships and taken 300 prisoners. A day or two later came the news that these New Englanders had embarked on Lake George in the ships they had captured, and were

Burgoyne's supplies cut off.

cutting off the last sources of supply. And now, while even on shortest rations there was barely three weeks' food for the army, Lincoln's main force appeared in front, thus swelling the numbers of the American army to more than 16,000. The case had become as desperate as that of the Athenians at Syracuse before their last dreadful battle in the harbour. So, after eighteen weary days, no word yet coming from Clinton, the gallant Burgoyne attempted, by a furious effort, to break through the lines of an army that now outnumbered him more than three to one.

On the morning of October 7th, leaving the rest of his army in camp, Burgoyne advanced with 1,500 picked men to turn the American left. Small as the force was, its quality was superb, and with it were the best commanders, — Phillips, Riedesel, Fraser, Balcarras, and Ackland. Such a compact force, so ably led, might manœuvre quickly. If, on sounding the American position on the left, they should find it too strong to be forced, they might swiftly retreat. At all events, the movement would cover a foraging party which Burgoyne had sent out, — and this was no small matter. Arnold, too, the fighting general, it was reported, held no command; and Gates was known to be a sluggard. Such thoughts may have helped to shape the conduct of the British commander on this critical morning. But the scheme was swiftly overturned. As the British came on, their right was suddenly attacked by Morgan, while the New England regulars with 3,000 New York mili-

Second battle at Freeman's Farm. Oct. 7; the British totally defeated by Arnold.

tia assailed them in front. After a short, sharp fight against overwhelming numbers, their whole line was broken, and Fraser sought to form a second line a little farther back and on the west border of Freeman's Farm, though the ranks were badly disordered and all their cannon were lost. At this moment, Arnold, who had been watching from the heights, saw that a well-directed blow might not only ruin this retreating column, but also shatter the whole British army. Quick as thought he sprang upon his horse, and galloped to the scene of action. He was greeted with deafening hurrahs, and the men, leaping with exultation at sight of their beloved commander, rushed upon Fraser's half-formed line. At the same moment, while Morgan was still pressing on the British right, one of his marksmen shot General Fraser, who fell, mortally wounded, just as Arnold charged with mad fury upon his line. The British, thus assailed in front and flank, were soon pushed off the field. Arnold next attacked Lord Balcarras, who had retired behind intrenchments at the north of Freeman's Farm; but finding the resistance here too strong, he swept by, and charged upon the Canadian auxiliaries, who occupied a position just north of Balcarras, and covered the left wing of Breymann's forces at the extreme right of the British camp. The Canadians soon fled, leaving Breymann uncovered; and Arnold forthwith rushed against Breymann on the left, just as Morgan, who had prolonged his flanking march, assailed him on the right. Breymann was slain and his force routed; the British right wing was crushed, and their whole position taken in reverse and made untenable. Just at this moment, a wounded German soldier, lying on the ground, took aim at Arnold, and slew his horse, while the ball passed through the general's left leg, that had been wounded at Quebec, and fractured the bone a little above the knee. As Arnold fell, one of his men rushed up to bayonet the wounded soldier who had shot him, when the prostrate general cried, "For God's sake, don't hurt him; he's a fine fellow!" The poor German was saved, and it has been well said that this was the hour when Benedict Arnold should have died. His fall and the gathering twilight stopped the progress of the battle, but the American victory was complete and decisive. Nothing was left for Burgoyne but to get the wreck of his army out of the way as quickly as possible, and the next day he did so, making a skilful retreat upon Saratoga, in the course of which, during a skirmish, his soldiers burned General Schuyler's princely country-house, with its barns and granaries.

As the British retreated, General Gates steadily closed in upon them with his overwhelming forces, which now numbered nearly 20,000. Gates — to give him due credit — knew how to be active after the victory, although, when fighting was going on, he was a general of sedentary habits. When Arnold rushed down, at the critical moment, to complete the victory of Saratoga, Gates sent out Major Armstrong to stop him. "Call back that fellow," said Gates, "or he will be doing something rash!" But the eager Arnold had outgalloped the messenger, and came back

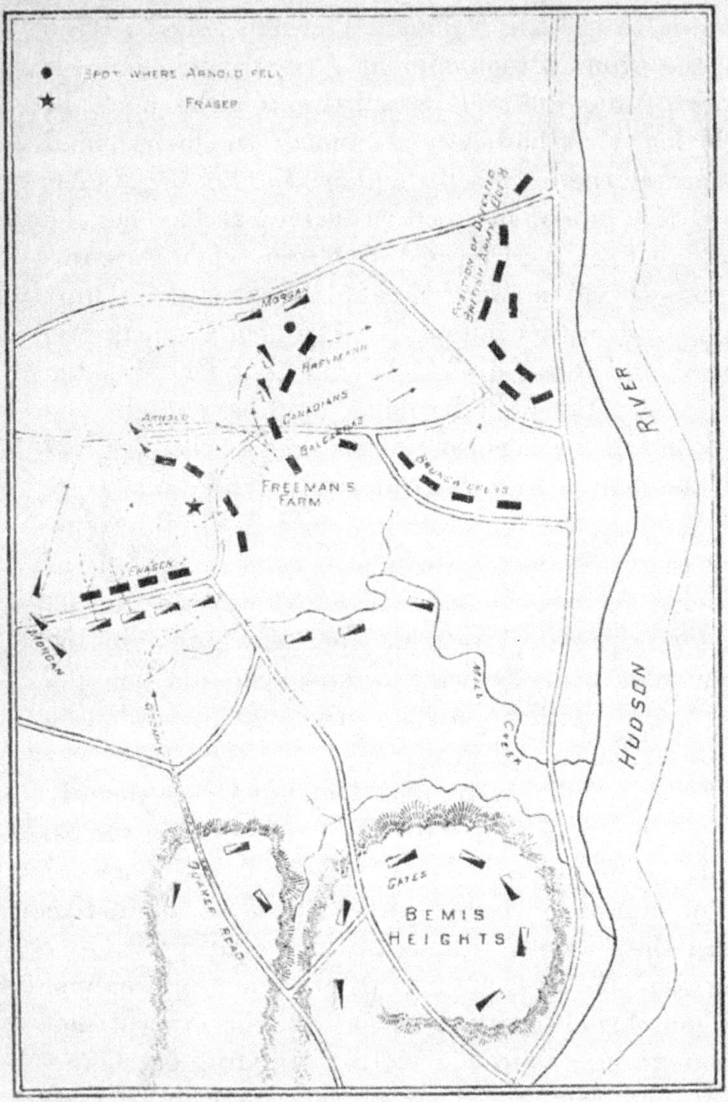

SECOND BATTLE AT FREEMAN'S FARM, CALLED BATTLE OF BEMIS HEIGHTS, OR OF STILLWATER

OCTOBER 7, 1777

only when his leg was broken and the victory won. In the mean time Gates sat at his headquarters, forgetful of the battle that was raging below, while he argued the merits of the American Revolution with a wounded British officer, Sir Francis Clarke, who had been brought in and laid upon the commander's bed to die. Losing his temper in the discussion, Gates called his adjutant, Wilkinson, out of the room, and asked him, "Did you ever hear so impudent a son of a b—h?" And this seems to have been all that the commanding general contributed to the crowning victory of Saratoga.

When Burgoyne reached the place where he had crossed the Hudson, he found a force of 3,000 Americans, with several batteries of cannon occupying the hills on the other side, so that it was now impossible to cross. A council of war decided to abandon all the artillery and baggage, push through the woods by night, and effect a crossing higher up, by Fort Edward, where the great river begins to be fordable. But no sooner had this plan been made than word was brought that the Americans were guarding all the fords, and had also planted detachments in a strong position to the northward, between Fort Edward and Fort George. The British army, in short, was surrounded. A brisk cannonade was opened upon it from the east and south, while Morgan's sharpshooters kept up a galling fire in the rear. Some of the women and wounded men were sent for safety to a large house in the neighbourhood, where they took refuge in the cellar; and

The British army is surrounded.

there the Baroness Riedesel tells us how she passed six dismal nights and days, crouching in a corner near the doorway, with her three little children clinging about her, while every now and then, with hideous crashing, a heavy cannon-ball passed through the room overhead. The cellar became crowded with crippled and dying men. But little food could be obtained, and the suffering from thirst was dreadful. It was only a few steps to the river, but every man who ventured out with a bucket was shot dead by Virginia rifles that never missed their aim. At last the brave wife of a British soldier volunteered to go; and thus the water was brought again and again, for the Americans would not fire at a woman.

And now, while Burgoyne's last ray of hope was dying, and while the veteran Phillips declared himself heartbroken at the misery which he could not relieve, where was Sir Henry Clinton? He had not thought it prudent to leave New York until after the arrival of 3,000 soldiers whom he expected from England. These men arrived on the 29th of September, but six days more elapsed before Sir Henry had taken them up the river and landed them near Putnam's headquarters at Peekskill. In a campaign of three days he outwitted that general, carried two of the forts after obstinate resistance, and compelled the Americans to abandon the others; and thus laid open the river so that British ships might go up to Albany. On the 8th of October, Sir Henry wrote to Burgoyne from Fort Montgomery: "*Nous y voici*, and nothing between us and Gates.

Clinton comes up the Hudson, but it is too late.

I sincerely hope this little success of ours will facilitate your operations." This dispatch was written on a scrap of very thin paper, and encased in an oval silver bullet, which opened with a tiny screw in the middle. Sir Henry then sent General Vaughan, with several frigates and the greater part of his force, to make all haste for Albany. As they passed up the river, the next day, they could not resist the temptation to land and set fire to the pretty village of Kingston, then the seat of the state legislature. George Clinton, governor of the state, just retreating from his able defence of the captured forts, hastened to protect the village, but came up only in time to see it in flames from one end to the other. Just then Sir Henry's messenger, as he skulked by the roadside, was caught and taken to the governor. He had been seen swallowing something, so they gave him an emetic, and obtained the silver bullet. The dispatch was read; the bearer was hanged to an apple-tree; and Burgoyne, weary with waiting for the news that never came, at last sent a flag of truce to General Gates, inquiring what terms of surrender would be accepted.

Gates first demanded an unconditional surrender, but on Burgoyne's indignant refusal he consented to make terms, and the more readily, no doubt, since he knew what had just happened in the Highlands, though his adversary did not. After three days of discussion the terms of surrender were agreed upon. Just as Burgoyne was about to sign the articles, a Tory made his way into camp with hearsay news that

Burgoyne surrenders, Oct. 17.

part of Clinton's army was approaching Albany. The subject was then anxiously reconsidered by the British officers, and an interesting discussion ensued as to whether they had so far pledged their faith to the surrender that they could not in honour draw back. The majority of the council decided that their faith was irrevocably pledged, and Burgoyne yielded to this opinion, though he did not share it, for he did not feel quite clear that the rumoured advance of Clinton could now avail to save him in any case. In this he was undoubtedly right. The American army, with its daily accretions of militia, had now grown to more than 20,000, and armed yeomanry were still pouring in by the hundred. A diversion threatened by less than 3,000 men, who were still more than fifty miles distant, could hardly have averted the doom of the British army. The only effect which it did produce was, perhaps, to work upon the timid Gates, and induce him to offer easy terms in order to hasten the surrender. On the 17th of October, accordingly, the articles were signed, exchanged, and put into execution. It was agreed that the British army should march out of camp with the honours of war, and pile their arms at an appointed place; they should then march through Massachusetts to Boston, from which port they might sail for Europe, it being understood that none of them should serve again in America during the war; all the officers might retain their small arms, and no one's private luggage should be searched or molested. At Burgoyne's earnest solicitation the American general consented that these proceed-

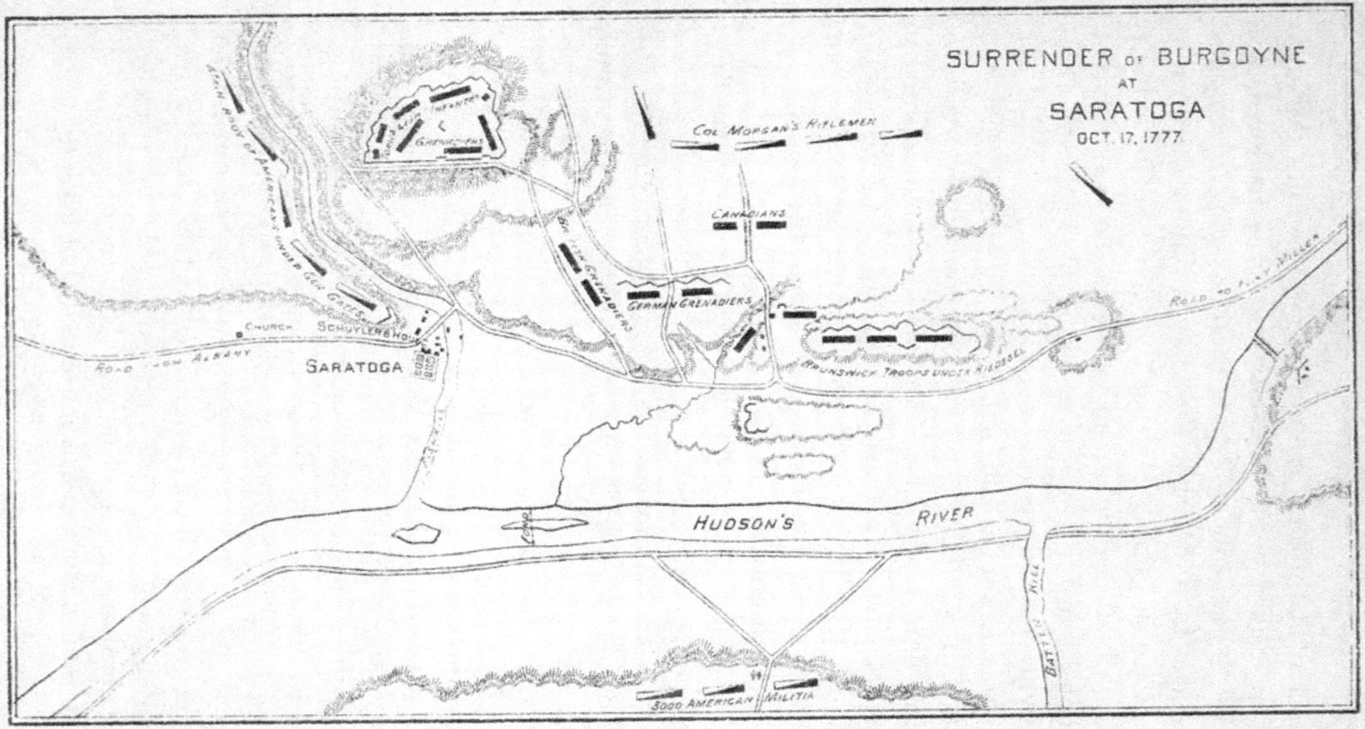
SURRENDER OF BURGOYNE
AT
SARATOGA
OCT. 17, 1777
COL MORGAN'S RIFLEMEN
CANADIANS
GERMAN GRENADIERS
BRUNSWICK TROOPS UNDER RIEDESEL
CHURCH
SCHUYLER'S HOU
SARATOGA
HUDSON'S
RIVER
3000 AMERICAN MILITIA

ings should be styled a "convention," instead of a surrender, in imitation of the famous Convention of Kloster-Seven, by which the Duke of Cumberland, twenty years before, had sought to save his feelings while losing his army, beleaguered by the French in Hanover. The soothing phrase has been well remembered by British historians, who to this day continue to speak of Burgoyne's surrender as the "Convention of Saratoga."

In carrying out the terms of the convention, both Gates and his soldiers showed praiseworthy delicacy. As the British marched off to a meadow by the river side and laid down their arms, the Americans remained within their lines, refusing to add to the humiliation of a gallant enemy by standing and looking on. As the disarmed soldiers then passed by the American lines, says Lieutenant Anbury, one of the captured officers, "I did not observe the least disrespect or even a taunting look, but all was mute astonishment and pity." Burgoyne stepped up and handed his sword to Gates, simply saying, "The fortune of war, General Gates, has made me your prisoner." The American general instantly returned the sword, replying, "I shall always be ready to testify that it has not been through any fault of your excellency." When Baron Riedesel had been presented to Gates and the other generals, he sent for his wife and children. Set free at last from the dreadful cellar, the baroness came with some trepidation into the enemy's camp; but the only look she saw upon any face was one of sympathy. "As I approached the tents," she says, "a noble-looking gentleman came toward me, and took the children out of the wagon; embraced and kissed them; and then, with tears in his eyes, helped me also to alight. . . . Presently he said, 'It may be embarrassing to you to dine with so many gentlemen. If you will come with your children to my tent, I will give you a frugal meal, but one that will at least be seasoned with good wishes.' 'Oh, sir,' I cried, 'you must surely be a husband and a father, since you show me so much kindness!' I then learned that it was General Schuyler."

Schuyler had indeed come, with unruffled soul, to look on while the fruit which he had sown, with the gallant aid of Stark and Herkimer, Arnold and Morgan, was plucked by an unworthy rival. He now met Burgoyne, who was naturally pained and embarrassed at the recollection of the beautiful house which his men had burned a few days before. In a speech in the House of Commons, some months later, Burgoyne told how Schuyler received him. "I expressed to General Schuyler," says Burgoyne, "my regret at the event which had happened, and the reasons which had occasioned it. He desired me to think no more of it, saying that the occasion justified it, according to the rules of war. . . . He did more: he sent an aide-de-camp to conduct me to Albany, in order, as he expressed it, to procure me better quarters than a stranger might be able to find. This gentleman conducted me to a very elegant house, and, to my great surprise, presented me to Mrs. Schuyler and her family; and in this gen-

Schuyler's magnanimity.

eral's house I remained during my whole stay at Albany, with a table of more than twenty covers for me and my friends, and every other possible demonstration of hospitality." Madame Riedesel was also invited to stay with the Schuylers; and when first she arrived in the house, one of her little girls exclaimed, "Oh, mamma! Is this the palace that papa was to have when he came to America?" As the Schuylers understood German, the baroness coloured, but all laughed pleasantly, and put her at ease.

With the generosity and delicacy thus shown alike by generals and soldiers, it is painful, though instructive, to contrast the coarseness and bad faith with which Congress proceeded to treat the captured army. The presence of the troops in and about Boston was felt to be a hardship, and General Heath, who commanded there, wrote to Washington, saying that if they were to stay till cold weather he hardly knew how to find shelter and fuel for them. Washington replied that they would not be likely to stay long, since it was clearly for Howe's interest to send them back to England as soon as possible, in order that they might replace other soldiers who would be sent over to America for the spring campaign. Congress caught up this suggestion with avidity, and put it to uses the furthest possible removed from Washington's meaning. When Sir William Howe proposed Newport as a point from which the soldiers might more speedily be shipped, Washington, for sound and obvious reasons, urged

Bad faith of Congress.

that there should be no departure from the strict letter of the convention. Congress forthwith not only acted upon this suggestion so far as to refuse Sir William Howe's request, but it went on gratuitously and absurdly to charge the British general with bad faith. It was hinted that he secretly intended to bring the troops to New York for immediate service, in defiance of the convention, and Congress proceeded to make this imputed treachery the ground for really false dealing on its own part. When Lord Howe's transports reached Boston, it was not only ordered that no troops should be allowed to embark until all the accounts for their subsistence should have been settled, but it was also required that these accounts should be liquidated in gold. In the instructions given to General Washington a year before, a refusal on the part of anybody to receive the Continental paper money was to be treated as a high misdemeanour. Now Congress refused to take its own money, which had depreciated till it was worth barely thirty cents on a dollar. The captured army was supplied with provisions and fuel that were paid for by General Heath with Continental paper, and now Congress insisted that General Burgoyne should make his repayment dollar for dollar in British gold, worth three times as much. In fairness to the delegates, we may admit that in all probability they did not realize the baseness of this conduct. They were no doubt misled by one of those wonderful bits of financial sophistry by which the enacting mind of our countrymen has so often been hopelessly confused. In an amusing letter to Washington, honest General Heath naively exclaims, "What an opinion must General Burgoyne have of the authority of these states, to suppose that his money would be received at any higher rate than our own in public payment! Such payment would at once be depreciating our currency with a witness." Washington was seriously annoyed and mortified by these vagaries, — the more so that he was at this very time endeavouring to arrange with Howe a general cartel for the exchange of prisoners; and he knew that the attempt to make thirty cents equal to a dollar would, as he said, "destroy the very idea of a cartel."

While these discussions were going on, Congress, like the wicked king in the fairy tale, anxious to impose conditions unlikely to be fulfilled, demanded that General Burgoyne should make out a descriptive list of all the officers and soldiers in his army, in order that if any of them should thereafter be found serving against the United States they might be punished accordingly. As no such provision was contained in the convention, upon the faith of which Burgoyne had surrendered, he naturally regarded the demand as insulting, and at first refused to comply with it. He afterwards yielded the point, in his eagerness to liberate his soldiers; but meanwhile, in a letter to Gates, he had incautiously let fall the expression, "The publick faith is broke [*sic*];" and this remark, coming to the ears of Congress, was immediately laid hold of as a pretext for repudiating the convention altogether. It was argued that Burgoyne had charged the United States with bad faith, in order to have an excuse for repudiating the convention on his own part; and on the 8th of January, Congress accordingly resolved, "that the embarkation of Lieutenant-General Burgoyne and the troops under his command be suspended till a distinct and explicit ratification of the Convention of Saratoga shall be properly notified by the court of Great Britain to Congress." Now as the British government could not give the required ratification without implicitly recognizing the independence of the United States, no further steps were taken in the matter, the "publick faith" was really broken, and the captured army was never sent home.

The behaviour of Congress was simply inexcusable.

In this wretched affair, Congress deliberately sacrificed principle to policy. It refused, on paltry pretexts, to carry out a solemn engagement which had been made by its accredited agent; and it did so simply through the fear that the British army might indirectly gain a possible reinforcement. Its conduct can be justified upon no grounds save such as would equally justify firing upon flags of truce. Nor can it be palliated even upon the lowest grounds of expediency, for, as it has been well said, "to a people struggling for political life the moral support derivable from the maintenance of honour and good faith was worth a dozen material victories." This sacrifice of principle to policy has served only to call down the condemnation of impartial historians, and to dim the lustre of the magnificent victory which the valour of our soldiers and the self-devotion of our people had won in the field. It was one out of many instances which show that, under any form of government, the moral sense of the governing body is likely to fall far below the highest moral standard recognized in the community.

The captured army was never sent home. The officers were treated as prisoners of war, and from time to time were exchanged. Burgoyne was allowed to go to England in the spring, and while still a prisoner on parole, he took his seat in Parliament, and became conspicuous among the defenders of the American cause. The troops were detained in the neighbourhood of Boston until the autumn of 1778, when they were all transferred to Charlottesville in Virginia. Here a rude village was built on the brow of a pleasant ridge of hills, and gardens were laid out and planted. Much kind assistance was rendered in all this work by Thomas Jefferson, who was then living close by, on his estate at Monticello, and did everything in his power to make things comfortable for soldiers and officers. Two years afterward, when Virginia became the seat of war, some of them were removed to Winchester in the Shenandoah valley, to Frederick in Maryland, and to Lancaster in Pennsylvania. Those who wished to return to Europe were exchanged or allowed to escape. The greater number, especially of the Germans, preferred to stay in this country and become American citizens. Before the end of 1783 they had dispersed in all directions.

What became of the captured army.

Such was the strange sequel of a campaign which, whether we consider the picturesqueness of its incidents or the magnitude of its results, was one of the most memorable in the history of mankind. Its varied scenes, framed in landscapes of grand and stirring beauty, had brought together such types of manhood as the feathered Mohawk sachem, the helmeted Brunswick dragoon, and the blue-frocked yeoman of New England, — types of ancient barbarism, of the militancy bequeathed from the Middle Ages, and of the industrial democracy that is to possess and control the future of the world. These men had mingled in a deadly struggle for the strategic centre of the Atlantic coast of North America, and now the fight had ended in the complete and overwhelming defeat of the forces of George III. Four years, indeed, — four years of sore distress and hope deferred, — were yet to pass before the fruits of this great victory could be gathered. The independence of the United States was not yet won; but the triumph at Saratoga set in motion a train of events from which the winning of independence was destined surely to follow.

Made in the USA
Monee, IL
04 November 2026

67387541R00213